Edited by
BAEK Seungman,
Jean MAGERAND, Claire BAILLY

ARCHITECTURE + LANDSCAPE + MEDIA
THE BIO DIGITAL CITY

CONTENTS

PREFACE

Following Palais de Tokyo, Paris-La Villette and Mons in Belgium, Bio-Digital City - the international event for the Future of city - is being held in Seoul, South Korea.
This exhibition takes us to a new level in Architecture and the Arts after long time efforts for a France-Korea collaboration. Not to mention the sponsorship of the Institut culturel Français, it is an achievement after years of attempts by the Ecole Nationale Supérieure d'Architecture de Montpellier, Ecole Nationale Supérieure d'Architecture de Paris-La Villette and Yeungnam University with the STAR program, to develop a government supported research exchange program between France and Korea. It is my great pleasure to witness Seoul being part of this creative movement.
The sponsorship was an inevitable result after a lot of anticipation from the French and Korean people. What would our future cities look like? What kind of urban planning could we adopt to design the centers of our societies which are full of aspirations for a better life for an increasing number of citizens?
The focus of this event lies on creativity; so participants from various fields can cooperate to produce diverse alternatives utilizing emerging technologies and to prove the necessity of sustainable development. This is evidence of how we are prepared to move towards a future that we have developed ourselves and also confirms our commitment towards a collaboration that will create positive results.
Starting at Palais de Tokyo in 2012, the 2014 international workshop in Seoul will let creative ideas drive the innovation and practicality of future cities while also considering the economic potentials. Smart grids, individualized supports, self-driving automobiles, diversification of new materials and even this very event itself is at the center of a creative economy.
What we have here is more than just an attempt because we will continually broaden the practical applications of emerging technologies and fundamental studies.
The Bio-Digital City Exhibition is a journey to specify the potential for positive contributions to urban areas in the future; and also an invitation from the innovative spirit to develop a creative economy.

파리 도쿄궁, 파리 국립과학관 그리고 벨기에 몽에 이어서, 서울에서 미래도시를 주제로 한 바이오디지털 도시(Bio-Digi-tal City) 국제행사를 개최합니다.

이번 전시회는 건축 예술분야에 있어서 프랑스와 한국간의 오랜 협력 끝에 맞는 새로운 단계입니다. 프랑스 문화원이 후원하는 이번 행사는 프랑스와 한국 정부의 지원을 통한 연구자 교환 프로그램(STAR program)의 혜택 속에 몽펠리에 국립건축대학, 파리 라 빌레뜨 국립건축대학, 영남대학교가 수년간의 협력을 통해 이뤄낸 결실입니다. 서울이 이러한 창조적인 움직임에 오늘날 함께하는 것을 보게 되어 매우 기쁩니다.

이번 후원은 양국민의 기대에 부응하기에 당연한 일이었습니다. 우리의 미래도시는 어떠할까요? 증가하는 시민수와 동시에 삶의 질 개선에 대한 열망으로 가득한 우리 사회의 중심부를 어떤 도시계획으로 디자인 할 수 있을까요?

다양한 분야의 참가자들이 상호 결합하여 신기술을 바탕으로 한 여러 대안들을 통합적으로 고안하고 지속 가능한 발전의 필요성을 증명해 보일 수 있게 할 수 있도록 주최 측에서는 본 행사의 중점을 창의성에 두었습니다. 이 행사는 우리 사회의 주도로 일궈진 미래를 향한 준비의 증거이자 결실을 맺고자 하는 의지의 증거입니다.

2012년 파리 도쿄 궁을 시작으로 2014년 서울 국제워크숍에서 발의되는 창의적 생각들은 혁신적이면서도 현실적 경제 잠재력을 고려한 미래도시 안을 그려볼 수 있도록 해줄 것입니다. 스마트그리드, 개별화된 지원, 자율주행시스템 차량 개발, 그리고 신소재의 다양화와 함께 여기 이 행사가 바로 창조경제의 주체가 될 것입니다.

금번 행사는 혁신적인 방법을 통해 신기술 및 기초연구 관련지식의 적용범위를 현실적으로 넓혀가며 구체화할 것이기에 단순한 시도 이상의 의미를 지닙니다.

이에 바이오디지털도시 전시회는 우리의 미래도시를 건설하는데 기여할 가능성들을 구체화하는 미래지향적 여정이 될 것이며, 혁신정신과 창조경제가 만들어 낸 미래로의 여행에 초대일 것입니다.

Daniel OLLIVIER

Directeur de l'Institut Français de Corée du Sud

주한 프랑스 문화원장

BEAK Seungman[1)]

A STUDY ON FUTURE CITY AND THE PROSPECT OF URBAN AGRICULTURAL ARCHITECTURE

In 2006, Seoul (Development) Institute inplemented the 'Seoul 100 years Plan' of futuristic urban planning for the city of Seoul in the next 100 years. The project was stopped at the stage of presentation for researches on future such as literature research, expert seminar, forum, etc. for 6 months, but it served as a momentum for domestic public organizations to plan the future city.2) In consideration of technical progress in today, the future city in 100 years could be just a utopian discourse with unforeseeable conclusion. However, it pursued to establish the most predictable development plan based on advanced cases, so that it had positive meaning to establish an individual future design based on urban identity beyond current issues.

In 2012, the School of Architecture in Yeungnam University signed the MOU for mutual exchange with ENSA Paris-La Villette, and at the same time, an international workshop of the Bio-Digital City begun to establish the future city design centered on advanced agriculture. Prof. Jean Magerand and Prof. Claire Bailly planned workshop for 10 days with subject 'What does the future hold for living and digital technologies in the city?' at l'Atelier international du Grand Paris (AiGP) in July, 2012.3) The international workshop of 2012 AiGP suggested various types of futuristic hyper-collective for 5,000 units such as 'peer to peer', 'agriculture oriented life', 'multiplication city', 'no street no shop', and 'mix city' with reflection of contemporary technology development and changing lifestyle in the suburb of Paris.

In July and August, 2013, five consecutive international workshops in order of Parametric, Automatization, Mutualization, Hybridization, Auto-organization with the subject of 'Digital City and Urban Agriculture Bio-high-tech' were held at Cité des Sciences et de l'Industrie, Universcience for six weeks, in which a new self-sufficient type of city organization centered on agriculture and deviated from current city organization centered on road network was in experiment. Particularly, as the Cité de Sciences et de l'Industrie, Universcience was the most popular public facilities drawing the number of visitors in Paris, the workshop was held at the concourse of the Cité de Sciences to open a forum for debate regarding the future city to the general visitors. Also, in Mons of Belgium as part of Festival VIA event, the international workshop with the same subject of 'Digital City and Urban Agriculture Bio-high-tech' was held and planned by Prof. Jean Magerand, Prof. Claire Bailly, and Prof. Etienne Godimus, in where a motion graphics exhibition of progressive Korean artists, titled as 'Bio-Digital Art & Architecture', was presented by Sookyoung Son (Art) and Seungman Baek (Architecture). In August of 2014, the international workshop was held at Seoul Citizen's Hall in Seoul City with the theme of 'Tomorrow' Seoul Project, Urban Agricultural Architecture'.

1) As an administrative coordinator and host of the international event in Seoul, 2014 Bio-Digital City, held as part of Korea-France international cooperation foundation project (STAR Program), he obtained diplôme (DPLG) and post-diplôme CEAA from ENSA Paris-Belleville and a doctorate from Ecole des Hautes Etudes en Sciences Sociales (EHESS) in Paris. And he has accumulated professional experiences in Seoul (Development) Institute, Junglim Architecture, and Space Group; currently, he is associate professor in the School of Architecture in Yeungnam University.

2) <Basic Research for Seoul 100 Plan>, Seoul Development Institute, February, 2007.

3) AiGP (Atelier International du Grand Paris) was established at Palais de Tokyo in 2008 to hold seminars, forums, workshops, and etc with international specialists in various fields such as architecture, city, landscape, and etc. to design the future of Paris.

A self-sufficient city in the outskirt of Paris, suggested at the Paris-La Villette Workshop 2013

2006년 서울(시정개발)연구원에서 100년 후의 서울을 위한 미래적 도시계획 '서울 100년 계획'을 진행하였다. 비록 예산부족으로 6개월간의 문헌조사, 전문가세미나, 토론회 등 미래연구에 대한 발제로 그쳤지만, 국내 공공기관이 미래도시를 계획하는 계기가 된다.2) 현대의 기술발전을 고려했을 때, 100년 후 미래도시는 예측불허의 결론이 없는 공상적인 담론에 그칠 수 있다. 그러나 선진사례를 통해 예측 가능한 발전계획을 수립하는 경향에 있어서, 현안문제를 넘어 도시정체성을 바탕으로 주체적인 미래구상을 하는 긍정적인 의미를 지닌다.

2012년 파리 라 빌레뜨 국립건축학교(ENSA Paris-La Villette)와 영남대학교 건축학부가 상호교류 MOU를 체결하는 동시에, 첨단농업을 중심으로 한 미래도시 구상인 The Bio-Digital City 국제워크숍이 시작되었다. 장 마젤랑(Jean MAGERAND) 교수와 끌레어 바이이(Claire BAILLY) 교수는 2012년 7월 그랑파리 국제아틀리에(AiGP)3)에서 'What does the future hold for living and digital technologies in the city?'라는 주제로 10일간의 워크숍을 기획하였다. 2012 그랑파리 국제워크숍에서는 파리외각에 현대 기술발전과 변화하는 생활상을 반영한 5,000세대 단위 미래형 고밀주거가 Peer to peer, Agriculture oriented life, Multiplication city, No street no shop, Mix city 등의 다양한 방법으로 제시되었다.

2013년 7-8월 파리 국립과학관(Cité des Sciences et de l'Industrie, Universcience)에서는 'Digital City and Urbain Agriculture Bio-high-tech' 주제로 6주에 걸친 5회 연속의 국제워크숍이 Parametric, Automatization, Mutualization, Hybridization, Auto-organization 순으로 진행되고, 도로망 중심으로 구성된 기존의 도시조직에서 벗어나 농업 중심의 자족적인 새로운 도시조직이 실험되었다. 특히 파리 국립과학관은 파리에서 가장 많은 관람객을 지닌 공공시설로서, 워크숍을 과학관 중앙홀에 개방하여 일반관람객에게 미래도시에 대한 토론의 장을 마련하였다. 또한 2014년 3월에 벨기에 몽에서도 Festival VIA 행사의 일환으로 'Digital City and Urban Agriculture Bio-high-tech' 동일주제의 국제워크숍이 장 마젤랑, 끌레어 바이이, 에띠엔느 고디무스(Etienne GODIMUS) 교수 기획으로 진행되었고, 진보적인 한국작가들의 'Bio-Digital Art & Architecture' 영상전이 손숙영(예술), 백승만(건축) 기획으로 이루어졌다.

그리고 2014년 8월에 서울시시민청에서 'Tomorrow's Seoul Project, Urban Agricultural Architecture' 주제로 The Bio-Digital City 서울 국제워크숍이 개최되기에 이른다.

1) 파리 벨빌 국립건축대학 디플롬(DPLG)/포스트디플롬(CEAA), 파리 사회과학고등연구원 박사를 취득하고, 서울(시정개발)연구원, 정림, 공간 등 실무를 거쳐 현재 영남대학교 건축학부 부교수이다.

2) <서울 100년 계획을 위한 기초연구>, 서울시정개발연구원, 2007.02.

3) 그랑파리 국제아틀리에(AiGP: Atelier International du Grand Paris)는 2008년 도쿄 궁(Palais de Tokyo)에 설립되어 건축, 도시, 조경 등 다양한 분야의 국제적인 전문가들과 함께 세미나, 토론회, 워크숍 등을 통해 파리시의 미래를 구상하고 있다.

AGRO-CITY SEOUL

Seoul City, where the population is 10 million, performed a grand-scale urban regeneration project, which included the recreation of downtown and the Han River Renaissance applying a culture city paradigm after Seoul's Cheonggyecheon Restoration Project (2003-2005), and was assigned as 2010.4) Creating higher value with the source of culture and shifting to 'culturenomics'5) according to government-leading culture city paradigm, recently, the paradigm of 'Agro City' was suggested to establish a type of cooperative village. Seoul City announced the 'ACT ON DEVELOPMENT AND SUPPORT OF URBAN AGRICULTURE' (Act of Urban Agriculture) on November 22, 2011, and declared 'the original year of urban agriculture in Seoul' on June 2, 2012. They also created a master plan for the urban agriculture (2014) in Seoul as well as various urban agriculture projects such as an edge vegetable garden, supplying box vegetable garden and making vegetable garden on rooftop and practical support for the urban agriculture in village communities, etc. The urban agriculture had various benefits for ageing society and unemployment issues to provide social enterprise, welfare and leisure. Wonsoon Park, the Mayor of Seoul, referred to the benefits of the urban agriculture at conference of the 4th Urban Agriculture Expo as "A program that has many positive effects. Not only can our citizens cultivate and eat crops by themselves, but also it can also help the broken city community, cultivate emotions and provide healing, improve education and health, provide leisure activities to the elderly and reduce greenhouse gases to play various roles demanded for making better city. Also, citizens will understand the farm village through the urban agriculture and become better mediators between the city and farm village, contributing to an urban–rural win-win. Furthermore, it is expected as future industry to create culture content and have a key part in creation of various occupations."6)

4) WDP (World Design Capital) was established by Peter Zec to develop the economy, enrich the culture, and improve the quality of citizen's life by design; and IDA (International Design Alliance) and ICSID (International Council of Societies of Industrial Design) make the system. Currently, Torino of Italy was assigned in 2008, Seoul of Korea in 2010, Helsinki of Finland in 2012, Cape Town of Republic of South Africa in 2014, and Taipei of the Republic of China in 2016.

5) The term of 'Culturenomics' is compound of culture and economics; it is initially suggested by Prof. Peter Duelund of University of Copenhagen in Denmark.

6) It was extracted from the welcome speech addressed at the conference of the 4th Seoul Urban Agriculture Expo held on June 5, 2015 at conference room located in the 3rd floor of the new city hall of Seoul City.

TOMORROW'S SEOUL PROJECT

The 2014 Seoul International Workshop placed emphasis on current practical possibilities in an extension of experimental research on the future Bio-Digital City.Thus, Seoul suggested the 'urban agriculture architecture' as a new architecture type to create a physical environment with regard to urban agriculture, which was one of the contemporary main subjects. The workshop targeted the city regeneration plan that was preceded in Dangsan-dong industrial district alongside the banks of the Han River, Yeouido business district, Ichon residential districts, and Apgujeong residential districts with consideration to the city landscape and connection of the urban agriculture architecture with the natural environment.

The Seoul International Workshop was a fusional meeting of architecture and media, standing for 'MediArchi' (Media + Architecture). It was performed under the joint direction of experts on architecture,·city and media, led by the architect of Kihong Kim. The result of each project was presented as 5-minutes video and 3D-printed models rather than panels or miniature. In conjunction with the workshop, the exhibition, titled 'Bio-Digital Art & Architecture', created by Korean artists, who participated in the Festival VIA motion graphics exhibition in Belgium, was planned and directed by the art director of Sookyoung Son. The opening seminar was presented by Jean Magerand, Claire Bailly, Hyunggi Kim (Chung Ang University), Ungyu Jang (Unsangdong), Taeseok Ha (SCALe) and Sehee Min (Randomwalks). With the subject of 'Fusion of nature and technology'. Finally with support of the French Cultural Center and direction by Marlène Vernouillet of Cité des Sciences et de l'Industrie (Universcience), the existing results of the international workshop for The Bio-Digital City were exhibited and explained to draw attention and help understanding of not only experts, but also the general public.

The master educator of the French Cultural Center in Korea, Yan Moreau, and the manager of Seoul Citizens Hall, Jeongpil, Choi, participated from the first stages of planning to the closing ceremony and gave the Award of French Ambassadors in Korea (A Tuft City), the Award of Seoul Foundation for Arts and Culture Director (Weave), and the Award of French Cultural Center Director in Korea (Striped Farming City) for the 3 artworks among the workshop results, chosen by professional judges. The awarded artworks as well as the 3D-printed models were donated to Cité des Sciences et de l'Industrie, Universcience.

아그로시티 서울

인구 1천만의 서울시는 청계천 복원사업(2003-2005) 이후 문화도시 패러다임을 대대적으로 실시하여 도심재창조, 한강 르네상스 등 대규모의 도시재생사업을 진행하고, 2010년 세계 디자인 수도(WDC) 4) 로 지정되었다. 문화를 원천으로 고부가가치를 창출하고 도시경쟁력을 높이고자 한 '컬쳐노믹스'(Culturenomics)5)의 정부주도형 문화도시 패러다임에 이어 최근에는 협동조합형 마을 만들기의 일환으로 '아그로 시티'(Agro City)가 제안되었다. 서울시는 2011년 11월 22일 '도시농업의 육성 및 지원에 관한 법률'(도시농업법)을 공포하고, 2012년 6월 2일 '서울 도시농업원년'을 선포한다. 그리고 서울시 도시농업 마스터플랜(2013)과 더불어 자투리텃밭 조성, 상자텃밭 보급, 옥상텃밭 조성, 마을공동체 도시농업 실천지원 등 다양한 도시농업 프로젝트를 추진하였다. 도시농업은 당면한 고령화 사회와 실업문제에 있어서 사회적 기업 및 복지, 여가 등을 제공하는 다양한 이점이 있다. 박원순 서울시장은 도시농업의 이점을 제4회 도시농업박람회 컨퍼런스에서 다음과 같이 언급하였다: "도시농업은 시민들이 농작물을 재배하여 직접 먹는 것 외에 다양한 긍정적 효과를 가지고 있습니다. 단절된 도시공동체 회복, 정서함량과 치유 효과, 교육 및 건강증진, 어르신들의 여가활동, 온실가스 감소 등 보다 살기 좋은 도시를 만들기 위해 필요한 다양한 역할을 수행하고 있습니다. 또한, 시민이 도시농업을 통해 농촌을 이해할 수 있도록 하여 도시와 농촌의 훌륭한 가교 역할을 해냄으로써 도농상생에 크게 기여하고, 더 나아가 미래 산업으로서 문화콘텐츠를 만들어 내고, 다양한 일자리 창출에도 일익을 담당할 수 있을 것으로 기대합니다."6)

4) 세계 디자인 수도(WDP: World Design Capital)는 디자인을 통해 경제를 발전시키고, 문화를 풍요롭게 함으로써, 시민 삶의 질을 개선하자는 취지로 페터 첵(Peter Zec)이 창안하였으며, 국제디자인연맹(IDA)과 국제산업디자인단체협의회(ICSID)가 제도를 만들었다. 현재 2008년 이탈리아 토리노, 2010년 대한민국 서울, 2012년 핀란드 헬싱키, 2014년 남아프리카 공화국 케이프타운, 2016년 중화민국 타이베이 시가 지정되었다.

5) '컬쳐노믹스'는 문화(Culture)와 경제(Economics)를 융합한 합성어로서 덴마크 코펜하겐대학 피터듀런드(Peter Duelund) 교수가 처음 제안했다.

6) 2015년 6월 5일, 서울시청 신청사 3층 대회의실에서 진행된 제4회 서울도시농업박람회 컨퍼런스의 환영사 중에서 발췌하였다.

내일의 서울 프로젝트

2014 서울 국제워크숍은 The Bio-Digital City 미래도시에 대한 실험적인 연구의 연장선상에서 현 시점에서의 실천 가능성에 주안점을 두었다. 따라서 서울시 현 중점과제의 하나인 도시농업을 보다 적극적인 방안으로 물리적 환경을 창출하는 새로운 건축유형으로서의 '도시농업건축'을 제안하였다. 그리고 도시농업건축의 자연환경과의 연계, 도시경관 등을 고려하여 도시재생계획이 진행 중인 한강변의 당산동 공업지구, 여의도 상업지구, 이촌 주거지구, 압구정 주거지구를 실험대상지로 워크숍이 진행되었다.

서울 국제워크숍은 특별히 건축과 미디어의 융합적 만남으로 'MediArchi'(Media + Architecture)를 표방하여 김기홍 건축가 중심으로 건축·도시 전문가와 미디어 전문가의 공동지도가 이루어졌으며, 각 프로젝트 결과물로서 기존의 판넬 및 모형이 아닌 5분 동영상과 3D 프린터 모델로 표현하였다. 워크숍과 동시에, 벨기에 Festival VIA 영상전에 참여했던 한국작가들의 'Bio-Digital Art & Architecture' 작품전시회가 손숙영 아트디렉터의 기획으로 마련되고, '자연과 기술의 융합'이라는 주제로 장 마젤랑과 끌레어 바이이, 김형기(중앙대), 장운규(Unsangdong), 하태석(SCALe), 최우람, 민세희(Randomwalks) 등의 공개 세미나가 이어졌다. 그리고 프랑스문화원 지원과 파리 라 빌레뜨 국립과학관의 마를렌 베르누이에(Marlène VERNOUILLET) 진행으로 기존의 The Bio-Digital City 국제워크숍 결과물들을 전시하고 설명함으로써, 전문가뿐만 아니라 일반시민의 관심과 이해를 돕고자 하였다.

주한 프랑스문화원의 얀 모로(Yan MOREAU) 교육관과 서울시시민청의 최정필 과장은 이번 행사의 기획단계에서부터 폐막식에 이르기까지 동참하여 워크숍 결과물 중 전문심사위원들에 의해 선정된 3개 작품들에 관하여 주한 프랑스대사상(A Tuft City), 서울문화재단장상(Weave), 주한 프랑스문화원장상(Striped Farming City)을 수여하였으며, 수상작품을 포함한 이번 워크숍에서의 3D 프린터 모델들은 모두 파리 국립과학관에 기증되었다.

URBAN AGRICULTURE ARCHITECTURAL AND VERTICAL FARM

Recently, the number of farmhouses and farmers in korea have sharply decreased but the speed of population tends to be higher as compared to other countries. Thus, combined agriculture,7) plant farms, and city plan for 100-years-old generation appeared as the 6th industry. Professor. Dickson Despommier of Columbia University, relayed that "the plant farm of 30-floor-scale can cover the foods for 50,000 people", presented the term of 'Vertical Farm (1999) and became generally known around the world. Afterwards, various types of vertical farms were suggested, but only a few such as the Sky Greens Farms in Singapore (2011), Local Garden in Vancouver of Canada (2012) and the vertical farm in Hope Town in Busan were established. Despite various benefits of the vertical farm such as establishing food infrastructure for food self-sufficiency, producing crops in areas where agriculture is impossible, reducing cost of transportation and unit cost of production, preventing forest destruction and the desertification phenomenon, there weresone problem including the initial high investment cost, costs for lighting and heating to reduce the energy efficiency and some resistance against the artificialness of not using soil but mostly LED light and hydroponics.

Even though the vertical farm by Dickson guaranteed productivity, it was not successful. However, his theory had impact on and applied to some cases. Most of them were recognized as experimental utopian architecture; recently, it has made as various appearances around the world. We call the new type of tendency in architecture 'Urban Agricultural Architecture'.8)

Urban Agricultural Architecture includes each element of urban agriculture and plant farms, but the vertical farm (plan factory) performs for the great productivity and is different from the urban agricultural (social program) performed in the exterior space of buildings. That is, the urban agricultural architecture actively applies the urban agriculture into the interior space of building and establishes mutual exchange relationship with residential spaces to emphasize the function of social community with abundant lifestyle for the residents. Accordingly, the urban agricultural architecture is not just a functional factory but directly adopts various benefits from the urban agriculture to the residential space in pursuit of mutual synergy effect as fusion architecture.

The types of urban agricultural architecture are categorized as a life type and a produce type according to the proportion of produced crops, and the life type is divided by private ownership type and joint ownership type according to the method of connection between the agricultural space and residential space.9) The life type is mainly for social benefits such as personal leisure, community, pleasant life environment, beautiful city landscape and to build a healthy food circulation system as a unit of architecture rather than to obtain a major source of income for the residents. Meanwhile, the produce type would become the major source of income for the residents, the agricultural space would be distinguished by a floor or building to be combined with the residential space, and a production facility system such as artificial lighting (LED), agricultural robot, u-IT (u-Farm) would be established. Cconnect the infrastructure of city to supply crops safely and immediately to the surrounding areas could also be considered. Therefore, the produce type is the most similar with the theory of the vertical farm of Dickson. In terms of connecting to the residential space, it could be various city architectures could be designed in accordance with the characteristics of places as deviated from uniform factory buildings only for production efficiency.

In regards to the city, the ultimate goal of the urban agricultural architecture is to build a 'self-sufficient city', as an ideal level of all kinds of urban planning. Generally, a new town project aiming to build the self-sufficient city so that creation of occupations for the residential population is the most important, but also the urban agricultural architecture satisfies to create the occupations related to agriculture and relevant industry so the theory of 'Neighbourhood Unit' could be applied to build desirable life environment. That is, the most ideal planning scale for the urban agricultural architecture would be the unit of 10,000~20,000 people (approximately 5,000 households) in the Neighbourhood Unit to establish the self-sufficient community of the residents with mutual cooperation. As for national and local competitiveness, the maximum volume of mass-production could be the goal of the advanced agricultural technology, but the advisable urban agricultural architecture at a global level is to be self-sufficient to minimize physical transportation between cities.

7) The 6th industry is multi-activity industry with creation of higher value by connecting and providing food or manufacturing, processing (2nd industry) and distribution, sales, culture, experience, tour, service (3rd industry), and etc. based on all tangible and intangible resources from farm village, led by the people of the farm village.

8) Taekwon Joung, Seungman Baek, <A Study on the Tendency Appearing Urban Agricultural Architecture for Green Life in the Future>, IJAER, no. 24, 2014.

9) Taekwon Joung, <A Study on the Tendency of Urban Agriculture Architecture and the Prospect>, Master's Dissertation of the School of Architecture in Yeungnam University, 2015.

도시농업건축과 수직농장

한국은 최근에 농가수와 농민수가 격감하는 반면에, 고령화 추세는 세계 어느 나라보다도 빠르게 진행되고 있다. 이에 6차 산업의로서의 복합농업7), 식물농장, 100세 시대의 도시계획 등이 나타나고 있다. 특히 식물농장은 "30층 규모의 빌딩농장이 5만 명의 식량을 책임질 수 있다"고 언급한 컬럼비아 대학의 딕슨 데스포미어(Dickson Despommier) 교수가 '수직농장'(The Vertical Farm, 1999)을 발표함으로써 세계적으로 널리 알려지게 되었다. 그 후, 수직농장은 다양한 형태로 많이 제안되었으나, 지어진 사례는 싱가포르 Sky Greens Farms(2011), 캐나다 벤쿠버의 Local Garden(2012), 부산의 희망마을 수직농장(2012) 등 아직 소수이다. 수직농장이 식량자급자족의 식량 인프라 구축, 농사불가능지역의 생산가능, 저렴한 운송비용과 생산단가, 숲 파괴 및 사막화 현상 방지 등 다양한 장점에도 불구하고, 초기 투자비용이 높으며, 조명과 난방 등의 비용이 높아 오히려 에너지효율이 떨어지는 등의 문제점이 지적되었다. 그리고 일부는 자연 채광과 흙이 아닌 대부분 LED 조명에 수경재배를 하는 인공성에 대한 거부감도 있는 듯하다.

비록 생산성을 담보한 딕슨의 수직농장이 현실적으로 성공하지 못하였으나, 그의 이론에 영향을 받아서 응용된 사례들을 찾아볼 수 있다. 이들은 대부분 실험적인 유토피아 건축으로 인식되곤 하지만, 최근에 다양한 양상으로 세계 곳곳에 실현되기 시작하였다. 우리는 이러한 새로운 경향의 건축유형을 '도시농업건축'(Urban Agricultural Architecture)이라고 명명하였다.8)

도시농업건축은 도시농업과 식물농장의 각 요소들을 포함하고 있으나, 건물의 외부공간에 실행하는 도시농업(사회적 프로그램)과 최대의 생산성을 위한 수직농장(식물공장)과는 차이점이 있다. 즉, 도시농업건축은 도시농업을 건물의 내부공간에 적극적으로 수용하고 거주공간과의 상호교류관계를 형성함으로써, 거주민의 풍부한 생활양식을 제공하는 사회적 커뮤니티 기능이 강조된다. 따라서 도시농업건축은 기능적 공장이 아닌 도시농업의 다양한 이점을 생활공간에 직접적으로 대입하여 상호 시너지효과를 추구하는 융복합 건축물이다.

도시농업건축의 유형은 농작물생산의 비중에 따라 크게 생활형(life type)과 생산형(produce type)으로 나뉘며, 그리고 생활형은 농업공간의 생활공간과의 결합방식에 따라 개인소유형과 공동소유형이 있다.9) 생활형은 주로 개인 여가활용, 공동체형성, 쾌적한 생활환경 및 아름다운 도시경관 조성 등의 사회적 이점을 위한 것으로 거주민의 주요 수입원이 되기보다는 건축물 단위로 건강한 먹거리 순환체계를 형성할 수 있다, 반면에, 생산형은 거주민의 주요 수입원이 될 것이기에, 효율적인 생산을 위해 농업공간이 층별 또는 동별로 구분하여 생활공간과 결합하고, 인공조명(LED), 농업로봇, u-IT(u-Farm) 등과 같은 생산설비시스템을 구축한다. 그리고 주변지역에 안전하고 신속한 농작물 공급을 위해, 도시 인프라와의 연계를 고려할 수 있다. 따라서 생산형은 딕슨의 수직농장 이론과 가장 근접한 것이지만, 생활공간과 결합하는 점에서 자칫 생산효율만을 위한 획일적인 공장건물에서 벗어나 장소의 특성에 따라 설계되는 다양한 도시건축이 될 수 있다.

도시차원에서, 도시농업건축의 궁극적인 목표는 모든 도시계획의 이상인 '자급자족 도시'를 만드는 것이다. 흔히 신도시 계획은 자족도시를 지향하여 정착인구를 위한 일자리 창출이 매우 중요한 고려사항이지만, 도시농업건축은 농업 및 관련 산업이 일자리 창출을 충족시킨 상황에서 바람직한 생활환경을 조성하기 위해 '근린주구'(Neighbourhood Unit) 이론을 접목할 수 있다. 즉, 도시농업건축의 이상적인 계획규모는 거주민들 간의 공동체를 형성하고 상호협동으로 자급자족할 수 있는 근린주구의 인구 10,000~20,000명(약 5,000세대) 단위가 될 것이다. 비록 국가적, 지역적 경쟁력 측면에선 첨단농업기술로 최대한 대량생산이 목표가 될 수 있을 것이지만, 범세계적인 차원에서 바람직한 도시농업건축은 도시간의 물리적 이동을 최소화하는 자급자족이다.

7) 6차 산업은 농촌주민이 중심이 되어 농촌에 존재하는 모든 유·무형의 자원을 바탕으로 식품 또는 제조, 가공(2차 산업) 및 유통, 판매, 문화, 체험, 관광, 서비스(3차 산업) 등을 복합적으로 연계, 제공함으로써, 부가가치를 창출하는 복합활동산업을 일컫는다.

8) Taekwon Joung, Seung-man Baek, <A Study on the tendency appearing Urban Agricultural Architecture for Green Life in the Future>, IJAER, no. 24, 2014.

9) 정태권, <도시농업건축의 경향과 전망에 관한 연구>, 영남대학교 건축학부 석사논문, 2015.

Korean buffets of the major companies spreading throughout the country(left)
A vertical farm of social cooperative (right, Busan)

An urban agriculture of the nursing home in the suburb of Daegu (2004, KIM Hyer-im)

PROSPECT OF URBAN AGRICULTURAL ARCHITECTURE

In Korea, the urban agriculture was diffused all around the country under the support of not only Seoul City, but also the Rural Development Administration as well as of local governments and nonprofit civic organizations and a 'local food' system was established to vitalize the local economy. In addition to public organizations and civic organizations, conglomerates engaged to widely expand the giant food chains of Korean cuisine buffets such as Nature Kitchen (E-Land Park), Season's Table (CJ Foodvill), Olga (Pulmuone), and etc. Meanwhile, the domestic vertical farms such as horizontal plant factory of Rural Development Administration (2004), container box type plant factory (2009), Iksan LED plant factory of LED-Agri-bio Fusion Technology Research Center in Chonbuk National University (2013) and Jecheon LED plant factory of Daekyung Plant Farm Promotion Agency in Gyeongbuk Technopark. were mostly in the stage of experimental research. The vertical farm in the Hope town in Sujeong-dong, Busan (2012) was only established in terms of urban regeneration to be managed by local village people and create social services and occupations for vulnerable people so that it was approved by the Ministry for Food, Agriculture, Forestry and Fisheries as the 1st 'vertical farm of social cooperative'.
Recently, They decided to build a 3-floor-scale vertical farm in a reconstruction apartment complex located in Mokdong, Seoul, with support of contributed acceptance instead of mitigation of capacity. In 2009, Namyangju-si, Gyeonggi-do tried to build 5~6-floor-scale of vertical farm on the ground with site of 300m2 for the first time in our country, but it was cancelled due to budget problems. The Ministry for Food, Agriculture, Forestry and Fisheries explained saying that "comparing the installation cost, it demanded 200~300 thousand won per pyeong (3.3m2) of vinyl greenhouse and 1 million won for glass greenhouse but a total 10 million won was necessary for building the vertical farm. However, the price of the produced crops could not make difference so it was difficult to perform the project". Today, Seoul City has low expectation with regard to its business value, but they would support the relevant technology development and obtain knowhow with respect to the future industry.10)

10) <The 'Vertical Farm' in Apartment... as Shown the First in Korea>, Journalist Seongwon Eum of Hankyoreh, April 14, 2015

However, in consideration of the speed changes in korea society and technology development, the development of vertical farms could be accelerated. In particular, if the conglomerates actively cooperated, such as the uniformed architecture culture Korean apartments or giant supermarkets it might be at risk of proceeding without connection with the members of local community and characteristics of the place. Particularly, considering the serious lack of systematic and design research for the vertical farm and applied architecture in the field of the domestic city and architecture, the low project feasibility is rather desirable in this situation.
In the 2014 Seoul workshop, it was aimed to graft the advanced technology and discussions on the urban agriculture (nature) pursued by the Bio-Digital City onto the domestic circumstances and experiments and diffuse it in terms of architecture and esthetics above all. These progressive discussions were performed at this event by Seoul Citizens Hall, and the necessity must be effective for not only large city but also depressed local societies for the city regeneration.

This Seoul international event could achieve good performance based upon voluntary cooperation of many people and various support of public organizations. First of all, I would like to thank the the French Embassy in Korea, the French Cultural Center in Korea, Seoul Foundation for Arts and Culture, Seoul Citizens Hall, and Cité des Sciences et de l'Industriel. Special thanks to tutors of the Suejeong Ko, Guyhong Kim, Dongjo Kim, Seokwon Kim, Junsik Moon, Sun Song, Na An, Junghyun Woo, Youngseok Lee, Jongeun Lee, Suho Jang, and Hyunyi Jung, the artists of the Hyunggi Kim, Gina Kim, Hwajin Oh, Byungchan Lee, and Uram Choe who participated in the artist exhibition, architects of the dmp, KyOsK, SCALe, The System LAB, and Unsangdong, TAFA who made the hiphop song for 'Bio-Digital City Seoul 2014', architect of Kihong Kim and Robox who was in charge of 3D printer, and the students of Jieun Kim, Youngjoo Nam, Hanna Shin, and Taekwon Jeong who studied in the School of Architecture in Yeungnam University and did all the work making posters, guidebooks, books and promoting and directing this event. Baek Seungman

도시농업건축의 전망

10) 〈아파트에 '수직농장'... 한국에서 첫 선〉, 한계레 사회 음성원 기자, 2015년 4월 14일.

한국에서 도시농업은 서울시뿐만 아니라 농촌진흥청의 지원 하에 지자체, 비영리시민단체 등과 더불어 전국적으로 확산되고, 지역경제 활성화를 위한 '로컬푸드'(local food) 시스템이 구축되고 있다. 공공기관 및 시민단체 이외에도, 대기업이 합류하여 한식뷔페인 자연별곡(이랜드파크), 계절밥상(CJ푸드빌), 올가(풀무원) 등 대형 외식체인점들이 널리 활성화되고 있다. 한편으로 수직농장의 국내현황은 농촌진흥청의 수평형 식물공장(2004)과 컨테이너박스형 식물공장(2009), 전북대학교 LED-녹생명융합기술연구센터의 익산 LED 식물공장(2013), 경북테크노파크 대경식물농장육성사업단의 제천 LED 식물공장(2013) 등 대부분 실험연구단계에 있으며, 부산의 수정동 희망마을 수직농장(2012)이 유일하게 도시재생차원에서 설립되고 지역 마을주민들의 경영과 취약계층의 사회서비스 및 일자리 창출을 실현하여 농림수산식품부의 제1호 '수직농장 사회적 협동조합'으로 인가를 받았다.
최근 서울 목동 재건축 아파트 단지에 용적률 완화 대신에 3층 규모의 수직농장이 기부채납 형식으로 지어질 예정으로 확정되었다. 2009년 경기도 남양주시가 국내 최초로 부지 300m2에 지상 5~6층 규모로 수직농장을 지으려고 시도하였으나, 예산문제로 무산되었다. 농림축산식품부는 "설치비를 비교하면, 비닐온실이 평(3.3m2)당 20만~30만원, 유리온실이 100만원인데, 수직농장은 1,000만원이 필요하다고 한다. 그런데 생산된 농작물 가격이 차이가 나지 않아 사업이 되기 어렵다"고 설명하였고, 오늘날 수직농장을 추진하는 서울시 입장 또한 현재로선 사업성에 대한 기대는 낮지만, 미래 산업으로서의 관련된 기술개발과 노하우 획득을 지원하는 입장으로 설명하고 있다.10)

그러나 우리의 급속한 사회변화와 기술발전의 속도를 고려하였을 때, 수직농장의 개발이 가속화될 조짐이 보인다. 특히 대기업의 적극적인 공조상황이 된다면, 한국의 아파트, 대형마트 등에서의 획일적인 건축문화처럼, 지역사회의 구성원 및 장소성과는 별개로 진행될 위험이 있다. 특히 국내의 도시 및 건축분야에서 수직농장 및 응용건축에 대한 계획적, 디자인적 연구가 매우 부족한 상황이므로, 현재의 낮은 사업성이 오히려 다행스럽게 여겨진다.
2014 서울워크숍에서는 The Bio-Digital City가 지향하고 있는 첨단기술과 도시농업(자연)에 대한 논의를 국내 상황과 접목하여 우선 건축과 미학차원에서 실험하고 확산하고자 하였다. 이러한 진보적인 논의가 이번 행사를 통해 서울시시민청에서 진행되었으며, 그 필요성은 대도시뿐만 아니라 침체된 지역사회의 도시재생에 매우 유효할 것으로 보인다.

이번 서울국제행사는 많은 분들의 자발적인 협력과 공공기관의 다양한 후원으로 좋은 성과를 가질 수 있었다. 우선 주한 프랑스대사관과 주한 프랑스문화원, 서울문화재단과 서울시시민청, 파리 국립과학관의 관계자 여러분들에 많은 감사의 말씀 전해드리며, 특별히 워크숍의 프로젝트들을 직접 지도해주신 고수정, 김기홍, 김동조, 김석원, 문준식, 송선, 안나, 우정현, 이영석, 이종은, 장수호, 정현이 튜터님들, 작가전시회에 참여해주신 김형기, 김진아, 오화진, 이병찬, 최우람의 예술가님들과 dmp, KyOsK, SCALe, The System LAB, Unsangdong의 건축가님들, 'Bio-Digital City Seoul 2014' 힙합노래를 만들어주신 TAFA, 3D 프린터를 담당해주신 김기홍 건축가님과 Robox 사, 그리고 이번 행사의 포스터, 가이드북, 책자 등의 제작 및 홍보, 진행에 이르기 모든 일을 함께 한 김지은, 남영주, 신한나, 정태권 영남대 건축학부 학생 여러분께 무한한 감사를 드린다. BSM

Jean MAGERAND
& Claire BAILLY

SEARCHING FOR A NEW ARCHITECTURAL AND URBAN MODERNITY

THE OBSOLESCENCE OF MODERN HERITAGE

After the heroic modernity of post World War II, which built dreams for rich and poor countries, a harmful modernity, full of disappointments and bad conscience has gradually imposed its presence all over the world. The New Man and his twin, modern man, invented by the cutting edge of the 20th century were cajoled into the 1960s. They are now being totally questioned. Le Corbusier's Standard Man, an icon of progress, has proven to be a selfish waster who destroys his environment. He neglects his fellow man, as well as ecosystems and future generations.

In this context of 'bad modernity' and especially during recent decades, the very purpose of objective knowledge has been questioned. Noble sciences, technologies and their natural children, Science & Tech, which had been shown for a century as tools for nature-conquest, as generators of progress and happiness, as mankind's saviors, have now gradually been relegated to the status of 'accomplices of the environmental deterioration'. The introduction of a dose of 'return to old values', driven by a 'politically-correct historicism' added to a hint of progressivism, seemed to be, for a whole generation, the ideal remedy against modernist excesses. To this, must be added the swarming questions generated by phenomena and events, critical of the contemporary context. Thus, climate change is seen as resulting from the globalist techno-economy, that is itself accused of destructive or self-destructive madness. Moreover, the biological revolution is bringing about bio-ethics which only complicate these worries.

At the same time, scientists, knowledgeable people and other experts have discovered the ins and outs of information sciences and of real time; they became aware of the technical and cultural consequences of this new situation. They have quickly realized that the new methods which emerged in this context are now influencing all societies; and that these methods accelerate the globalization phenomena that permeates the fields of both economics and culture.

Any sudden change immediately induces strong resistance. In particular, information control and its tools, such as computers, were suspected of having a bad influence on individuals and their collectives. In addition, economic crisis spread fear among populations. They are a constant threat - when they are not effective. This atmosphere creates 'over-pressure' which immediately induces 'perceptual depression' effects and changes the way we look at our environment, our culture, our habits, and our modernity. Within this mutation of our icons and our societal values, the perception of the relationship between humanity and environment is suddenly reversed. The boundless looking planet has turned into a jar in just a few decades.

In this context of value crisis, of forced transformation in our biological environment, of undigested scientific and technological changes, a climate of blurred tension is setting up and is blurring the stakes in the debate by hiding causes and consequences. Understanding these phenomena is not simple because of moving context – which happens in an unusual technical world - generates technophobia and technophilia clashing in a duel/debate that is passionate, doctrinal, superficial and full of misunderstanding. Critical postures are replaced by principled stands. Therefore our collective neuroses disrupts the objectivity we would wish to keep on our contemporary world. All these apriorisms and contradictory discourses make it difficult to get the 'air of time' and an objective judgment on the relevance of conceptualising thoughts from our time.

eneral view of the Hy-
ercollective city (250 000
habitants)

BAILLY / J.MAGERAND with :
FERS, C.BLEMAND, S.PASQUIER,
BERNHARDT, P.BERTIN / 2011

근대 유산의 노후화

세계 2차 대전 이후에 부유한 나라와 가난한 나라 모두에게 꿈을 심어주었던 구제자적인 근대성 다음으로는 전 세계에 점차 실망과 불신으로 가득 찬 해로운 근대성이 도래하였다. 20세기 전위운동(avant-garde)에 의해 도입되었던 '새로운 사람', '근대적 사람'은 1960년대까지는 총애를 받았다. 그러나 그들은 오늘날 완전히 문제가 되고 있다. 진보의 아이콘으로 여겨졌던 르 꼬르뷔지에의 '표준적인 사람'은 그의 환경을 파괴하는 이기적인 낭비자로 증명되었다. 그의 동료들은 물론, 생태계와 미래세대도 저버리고 있다.

'나쁜 근대성'의 맥락에서, 특히 최근 몇 십년동안 객관적 지식의 궁극성 또한 문제시 되어왔다. 숭고한 과학, 기술, 그리고 그 부산물인 과학기술은 지난 세기동안 진보와 행복을 만들어 내는 자연정복의 기계 또는 인류의 구원자처럼 인식되었으나, 점차 '환경파괴의 공범자'로 전락되었다. '옛 가치로의 회귀'를 소개하는 것은 진보주의가 살짝 가미된 '정치적으로 문제가 없는 역사주의'에 의해 진행되는 것인데, 이는 모든 세대에게 있어서 지나친 근대주의자에 대한 이상적인 처방처럼 보인다. 여기에 현시대의 맥락에 있어서 어떤 현상, 사건, 결정요인에 의해 만들어진 수많은 의문점들을 추가해야한다. 또한 기후변화는 자기파괴의 광기에 기인한 세계주의적 기술경제의 결과처럼 보인다. 더 나아가 생물학적 혁명은 이러한 의문들을 오히려 복잡하게 하는 생명윤리의 출현을 이끌어낸다.

이와 동시에, 과학자, 지식인 그리고 많은 전문가들이 정보과학과 실시간에 대한 많은 것을 밝혀내고 있다. 그들은 새로운 국면의 기술적, 문화적 결과들에 대해 자각하고 있다, 즉 이러한 맥락에서 새로운 방법론들이 경제와 문화 모두에 세계화 현상을 가속화시키는 사회적 영향을 바로 인식하였다.

급작스러운 모든 변화는 오히려 강한 망설임을 가져온다. 특히 정보기술과 컴퓨터는 개인과 공동체에 나쁜 영향을 미치는 것으로 의심을 받았다. 게다가 실효적이지 않을 때에는 경제적 위기가 사람들에게 두려움을 주고 지속적으로 위협한다. 이러한 사회적 분위기는 오히려 우리의 환경, 문화, 관습, 근대성에 관한 관점을 변화시키면서 '지각적 쇠퇴'를 가져오는 '중압감'을 만들어낸다. 이는 우리의 아이콘과 사회적 가치의 변이과정에서 인류와 환경 간의 관계에 대한 갑자기 역전된 관점이다. 광대한 행성이 몇 십 년 전에 하나의 병 규모가 되었다.

이러한 가치위기와 더불어 자연스럽지 않았던 생물학적 환경의 변화, 그리고 사람들이 미처 따라잡지 못한 과학적, 기술적 변화들의 맥락에서, 긴장을 흐려놓는 움직임이 생겨나고 있다. 이는 원인과 결과를 명확히 볼 수 없도록 하여 논의의 심각성을 흐려놓는 것이다. 이러한 현상을 이해하는 것은 상황의 맥락 자체가 전례없는 기술사회 속에서 계속 변화하기 때문에 쉽지 않은 일이다. 이 기술사회는 기술포비아(기술에 대한 공포와 두려움)와 기술필리아(기술에 대한 비이성적 맹신)를 만들어 내며, 이들이 격렬하면서도, 이론뿐인, 겉핥기식의 논쟁을 하도록 한다. 이러한 논쟁은 심지어 오해가 대부분인 경우도 많다. 비판의 자세는 원칙주의적인 견지로 교체된다. 따라서 우리의 집단 신경증은 우리가 현시대에서 지키고자 하는 객관적 이성을 방해하는 것이다. 이 모든 아포리즘과 모순되는 담론들은 '시대의 분위기'(air of time)를 이해하고 우리 시대의 사상들을 개념화할 수 있는 객관적인 판단을 어렵게 한다.

Our goal in this questioning world is to be detached enough to discern better the incidences of this complex context on the evolution of thoughts, speeches, theories, practices and productions related to the living environment. In order to find marks in this ambiguous situation, we can rely on history, the one which focuses on mechanisms, not only on the symbolism of facts. This history teaches us that this situation of big questioning is recurring in the evolution of cultures and of civilizations. In certain transitional periods, totally contradictory speeches coexist in a strong way. An a posteriori evaluation shows that permanent signs live in these particular times and that it is possible to find elements of constancy susceptible to enlighten us on our present situation. We are looking for these signs and their properties. The digital revolution and the revolution of knowledge of the living mechanisms take us deep into one of those critical moments. In our opinion, this is where the concept of Bio-digital emerges from.

THE GENES OF A MUTANT CULTURE

Rejection of techno-science is obviously not sufficient to re-format the 'air of time' or even the direction of history. In our context of 'declining modernity', agriculture, city, and the IT revolution are considered by dominant discourses as responsible for the destruction of the planet and as the cause of mental and physical ill-being of society. These doctrines come from a prevention-based approach, but they do not carry any rational change process. It is therefore very important to discern what is emerging and what is disappearing in our contemporary civilization.

Everything shows that today, in order to avoid long term self-destruction of the planet and preserve life on earth, it is necessary to better highlight potential tracks for a new civilizational construction. To learn more about our time, we must describe and assess the active agents of current changes.

For example, currently, the evolution of reality analysis modes, based on ecosystem models, is amplified by new analysis tools born from the méthodico-technical changes. By forming this way, the new context of life and thought acts directly on the way we analyze reality and the way we approach and understand it.

Today, a chain reaction is starting, generating artificial hyper-readings on a over-complex reality. This way, our artificial environment is re-presented and re-constituted. It changes the way we perceive it and perceive ourselves within it. The question of the numerous consequences of these new readings on our way to re-write the reality, to create it, immediately raises. The same question arises for all kinds of mechanisms, devices, and new processes that are taking place in the wake of new technologies and their new methodological universes.

It is once more informed on these new aspects of our time that we will be able to build useful tools to better understand the mechanisms that govern the influence of the latest scientific approaches of reality into other fields such as those of conception.

This 'air of time', that we try to introduce in design, was, for example, considered a 'inevitable formatting' by promoters of the Modern Movement. It totally depended, in their mind, on a multitude of theoretical, conceptual, material, immaterial, natural or artifactual facts constituting the vanguards. To them, all these heterogeneous specificities were guided by the compass of a perfectly identifiable 'progress', and were associating and hybridizing with each other, generating a common framework of thinking and living that gave it substance.

Today, Edgar Morin's work (La nature de la nature) or Joël de Rosnay's (Symbiotic Man) suggests to us a 'deep' air of time where units, multitudes, multiplicities, systems, equipment, beliefs, and ambiguities are mixed in a complexity which is rational but inaccessible in its overall coherence. They open to universes that which can be evocated but not fathomed in their entirety. According to these new philosophies, this constitutive realities imbricated within each other and making the major reality of our complex environment are approached under a specific global angle. The purpose of these new re-compositions of reality is to highlight better the organizations of these co-existing realities and to understand better the mechanisms governing their activities, their relationships, their development and their ability to act.

이러한 실정에서 우리의 목적은 우리 거주 환경에 대한 생각과 연설들, 이론, 관행과 생산의 복잡성을 좀더 잘 포착할 수 있도록 충분히 거리를 두는 것이다. 애매모호한 상황에서 표지(標識)를 찾기 위해서는, 사실의 상징성은 물론, 메카니즘에 초점을 두는 역사에서 실마리를 찾을 수 있다. 역사는 우리에게 이러한 의문의 시대가 문화와 문명의 진화에서 반복되었음을 가르쳐 준다. 특정 과도기에서는 완전히 모순되는 의견들이 강력한 목소리로 공존하곤 한다. 귀납적인 평가는 현대의 지속적인 징후들이 현시대에 나타나고 있으며, 이것이 우리의 현재에 대한 깨달음을 줄 수 있다. 우리는 이러한 징후와 그 속성들을 찾고 있다. 디지털 혁명과 삶의 메카니즘에 대한 지식 혁명은 우리를 결정적인 순간으로 깊숙이 이끌어 간다. 이것이 바로 '바이오 디지털'의 개념이 출현하게 된 계기이다.

돌연변이 유전자 문화

기술과학을 무조건적으로 거부하는 것은 분명히 '시대의 분위기'(air of time), 혹은 역사의 방향을 재구성하기에 충분하지 않다. '쇠퇴하는 근대성'의 맥락에서 농업, 도시, IT 혁명은 지배적인 담론으로 여겨지며, 이 행성의 파괴와 사회의 정신적, 물리적 피폐의 원인으로 보여지고 있다. 이러한 사고는 예방중심의 접근에서 오지만, 이는 아무런 이성적인 변화 과정을 수반하고 있지 못하다. 그러므로 현 문명사회에서 무엇이 나타나고 무엇이 사라지는 지를 포착하는 것이 매우 중요하다.

장기적으로 지구의 자멸을 피하고, 생명체를 보존하기 위하여, 새로운 문명 건설을 위한 잠재적인 길을 닦는 것에 더 많은 관심을 기울여야 한다. 현 시대에 대해 좀더 잘 이해하기 위하여, 우리는 현재 일어나고 있는 변화들의 조짐을 면밀히 살피고 평가해야 하는 것이다.

예를 들어, 생태계 모델에 기반하여, 현재 실재상황분석 모드의 진화는 의료기술 변화에서 만들어진 새로운 분석툴에 의하여 증폭되었다. 이러한 접근에서, 삶의 새로운 맥락과 생각들은 현실을 분석하고, 접근하며, 이해하는 방식에 직접적으로 영향을 미친다.

오늘날 지나치게 복잡해진 현실에 대해 인위적으로 과도한 분석이 만들어 지면서, 특정한 연쇄반응이 시작되고 있다. 이런 현상이 계속된다면, 우리의 인위적인 환경은 재게시되며 재구성될 것이다. 결과적으로 우리가 환경을 받아들이고 우리 자신을 그 안에서 받아들이는 방식마저 바꾸게 된다. 문제는, 우리가 현실을 다시 쓰고, 창조하는 방식에 대한 새로운 분석으로 인한 수없이 많은 결과가 곧바로 나타날 것이라는 점이다. 동일한 의문점이 새로운 기술과 새로운 방법론적 우주에 뒤이어 나타나는 메카니즘, 기기들, 새로운 프로세스들에 대하여 제기될 것이다.

우리가 유용한 툴을 사용하여 현실과 그 개념에 대한 최신 과학적 접근이 미치는 영향의 메커니즘을 보다 잘 이해할 수 있게 된다는 현시대의 새로운 국면이 좀더 알려졌다.

이러한 '시대의 분위기'는 우리가 설계를 통해 소개하고자 하는, 예를 들어 근대 운동의 촉진자들에 의해 '불가피한 형태구축'으로 생각된다. 이는 그들의 생각 속에서 선봉을 이루는 이론, 컨셉, 물질적, 비물질적, 자연적 혹은 인위적인 사실들에 완전히 의지된다. 그들에게 있어 서로 다른 종류로 이루어진 특수함은 완벽하게 인식 가능한 '진보'의 나침반에 의해서 안내되며, 서로 연결되고 융합되어 실체를 만들어내는 공통의 프레임을 구성한다.

오늘날, 에드가 모린의 작품(La nature de la nature) 혹은 조엘 드 로즈네이의 작품(공생 인간)은 우리에게 '깊은' 시대적 분위기을 제시하는데, 이는 단위와 군중, 다양성, 시스템, 장치들, 신념, 그리고 애매모호성이 모두 한데 복잡성으로 섞여서 이성적이나 접근 불가능한 일관성을 이루고 있다는 것이다. 이들은 존재를 일깨울 수 있으나 전체를 가늠할 수는 없는 우주로 열려있다. 이러한 새로운 철학들에 따르면, 이 실존의 구성은 서로 겹겹이 겹쳐서 우리의 복잡한 환경의 중심을 만들고 특정한 세계적 관점에서 접근할 수 있다. 이러한 새로운 재구성의 목적은 공존하는 현실들의 구성을 보다 분명하게 강조하고 그 활동과 관계, 발전과 행동할 수 있는 능력을 좌우하는 메커니즘을 보다 잘 이해하는 데에 있다.

Today, some great revolutionary paradigmatic phenomena are structuring the contemporary context; they look very promising in terms of change. Thus, information sciences have committed a technical transformation which itself causes, directly and indirectly, behavioral revolutions, because, in particular, of the exponential nature of the flow of information. The setup of real time, more and more present and pressing day after day, actually allows the fight against the pernicious effects of globalization phenomena which it nervertheless helped to generate. Complexity Sciences sharpen and amplify our understanding of reality by generating an intellectual revolution. The new awareness of our surroundings - boosted by the sciences of complexity - also helps to re-format and to mutate our views on the environment. We are able to think differently about our actions within the global ecosystem. Biology is carrying critical issues, it comes on the forefront and proposes new approaches to living beings, which become less magical but more amazing. It certainly holds great fears but also huge hopes.
Digital seeps into the production of knowledge, generating the same worries and the same fascinations. A new consciousness and a cultural revolution across the globe, of an unprecedented magnitude in history, slowly seem to emerge. All this new data, forming civilizational entities, agglomerate and/or mix and/or make synergies to influence the evolution of humanity and its knowledge. Together, they engage this major transformation of our relationships with our societal, conceptual, and socio-cultural environments. In our time, as in all other transition periods already experienced by humanity, the nature and the relative positions of each of these human variables vary within this context.
The way we look at them is also changing. As in any system, when viewed from the perspective of its ecosystem structure, context formats each of its sub-parts. A more or less active and effective hybridization is set up according to place and moment. In return, each entity reacts synergistically or in contradiction with the general trends; it participates in the changing context. So forced by the system, it operates by analogy, reaction, inertia or feedback; it mutates by adapting itself. It is in this very new and very particular context that our lifestyles and our cultures are likely to change along with our requirements and/or our needs of new life settings, better adapted to new situations that we built on our own.

SEARCHING FOR A WAY OF LIFE FOR A NEW MODERNITY

Because it is aware of these contemporary reluctances and of the complex morphology of the new civilization and of philosophical changes, the International Experimental Studio for Bio-Digital City has been continuing its research for several years. This research is about the impacts of this 'new deal' on our living places, that is to say on our planet as a living place. What will our natural areas, our landscapes, our agricultural systems, our architectures, our cities, and our societies, be tomorrow? How will we live, work, and collaborate? How can we rethink our ways of doing things to stop wasting territories, natural resources, and heritages? How can we get rid of the suicidal globalization based on competition and antagonism? How can we develop more civilized societies, that would be less selfish, less wasteful, more inclusive, more tolerant, more collaborative, more based on mutual help, partnership, sharing, solidarity, and mutualisation? Obviously, these qualities are not only carrying their own ethics, but they also appear as possible keys to invert the slow self-destructive process in which human organizations have ventured for one and a half centuries.

One of our working hypotheses is that this cultural revolution, including ethical values, is a must in which digital tools occupy a strategic position. They have indeed become a common denominator between people massively using digital technology for everyday life and ventures, which are forced for their development to acquire new skills on the subject every day. In our opinion, the general public is not yet convinced but they are aware of the high-tech research that aims at environmental improvements through the scale of the flat (home automation), of the building (building automation), of the district (smart-city), of the country (smart transportation), and worldwide (sustainable development assisted by satellite). These techniques can be applied to urban or agricultural land or can hybridize with other existing know-how, to provide new techniques of improvement of living conditions in the city and in the countryside.

오늘날, 몇몇 훌륭한 혁명적이고 모범적이라 할 수 있는 현상은 현시대의 맥락을 구조화하는 것이다. 이는 변화의 관점에서 매우 좋은 움직임이다. 따라서 정보과학은 기술적인 변화에 전념해왔는데, 이는 특히 폭발적으로 증가하는 정보의 흐름에 의해서 직간접적으로 행동의 혁신을 불러일으키는 것이다. 이러한 실시간 구성은 날이 갈수록 더욱 현실속으로 들어오고 있는데, 실제적으로 세계화 현상을 만들어내는 동시에 이에 대립할 수 있도록 하기도 한다. 복잡성의 과학은 지적 혁명을 만들어냄으로써 현실에 대한 우리의 이해를 더욱 날카롭게 하고 증폭시킨다. 복잡성의 과학으로부터 힘을 받은 우리 주변 환경에 대한 새로운 인식은 환경에 대한 우리의 관점을 다시 구성하고 돌연변이처럼 변화시키는 것을 돕는다. 우리는 전세계 생태계적 관점에서 우리의 행동에 대하여 다르게 생각해 볼 수 있다. 생물학은 생물에 대하여 새로운 접근을 제안하며 심각한 이슈들을 제기하고 있다. 이는 커다란 공포와 이에 맞먹는 큰 희망을 동시에 품고 있는 것과 같다.

디지털은 지식의 생산에 스며들어, 동일한 우려와 매혹을 만들어 내고 있다. 전세계적으로 일어나는 새로운 의식과 문화적 혁명은 역사적으로 그 전례가 없는 규모로 서서히 진행되고 있다. 이 모든 데이터는 문명의 실체를 구성하며, 한데 뭉치거나 섞여들어 시너지를 만들어 인류 진화 혹은 그 지식에 영향을 준다. 이들은 한데 어울려 우리와 우리 사회, 개념적, 사회문화적 환경과의 관계에 주요한 변형을 일으킨다. 우리의 시대에서, 자연과 인간 변수의 상대적인 위치는 맥락 속에서 변화한다.

우리가 그들을 바라보는 관점도 역시 변화한다. 여느 시스템과 마찬가지로, 생태계적 구조 관점에서 바라보면, 맥락은 각각의 부분집합을 구성한다. 다소간 차이는 있을 수 있으나, 적극적이고 효과적인 교배가 장소와 시간에 따라 형성된다. 그 대신에, 각 존재는 시너지를 만드는 방향으로 반응하며, 혹은 일반적인 트렌드에 모순되는 반응을 보인다. 이는 맥락을 변화시키는데에 참여하게 된다. 따라서 시스템에 의해 강요되면, 이는 비유, 반응, 관성 혹은 피드백에 의해 운영되며, 그 스스로에 적응해 감으로써 돌연변이를 만들어 낸다. 이는 굉장히 새롭고 특정한 맥락, 즉 우리의 라이프스타일과 문화가 우리의 필요요건에 따라서 변화하는, 혹은 새로운 삶의 방식에 대한 우리의 필요에 속하는데, 우리 스스로 만들어 나가는 새로운 상황에 보다 잘 맞아들어간다.

새로운 근대성을 위한 새로운 삶의 방식 탐색

현시대의 반항과 새로운 문명과 철학적 변화에 대한 복잡한 형태학을 인식하고 있기에, 바이오 디지털 시티를 위한 국제 실험 스튜디오는 수년간 연구를 계속해 오고 있다. 이 연구들은 우리 거주지역의 '대변혁'(new deal)에 대한 영향을 주제로 하고 있는데, 바로 지구 전체를 거주지역으로 보는 관점이다. 우리의 자연지대, 풍경, 농업시스템, 건축물, 도시들, 사회 그리고 내일은 어떤 것이 될까? 우리는 어떻게 살아가고, 일하고 협업해 나가게 될까? 어떻게 대지와 자연 자원, 그리고 우리에게 남겨진 유산을 더 이상 낭비하지 않도록 방법을 다시 생각해 볼 수 있을까? 어떻게 하면 자멸에 가까운, 경쟁과 적대감에 기초한 세계화를 없앨 수 있을까? 어떻게 좀더 문명화된 사회, 즉 이기적이지 않고, 낭비되지 않고, 보다 포용력이 높고, 보다 협력적이고, 상호간의 도움에 기초하며, 파트너쉽이 있고, 서로 공유하고, 연대감을 느낄 수 있고, 상호적인 그런 곳으로 만들 수 있을까? 의심의 여지 없이 이러한 가치들은 단순히 그 윤리를 품고 있을 뿐만이 아니라, 인간의 조직이 한세기 반에 걸쳐 진행시킨 자멸의 과정을 더디게 만들 수 있는 가능성의 핵심이다.

우리의 진행되는 가설들 중 하나는 바로 윤리적 가치를 포함하는 이러한 문화적 혁명이 디지털 도구들의 전략적 위치 차지에 있어서 필수적이라는 것이다. 이는 확실히 매일의 삶속에서 디지털 기술을 사용하는 사람들 사이에서 공통분모가 된다. 발전을 강요받는 벤처들은 새로운 기술과 주제들을 매일 습득하게 된다. 일반 대중은 아직 그들이 이미 환경 개선을 위한 첨단 기술 연구를 이미 인식하고 있다는 것에 그다지 설득되지 않았다. 그 규모는 아파트(가정 자동화), 혹은 빌딩단위(빌딩 자동화), 지역단위(스마트 시티), 국가(스마트 교통) 그리고 전 세계단위(인공위성이 지원하는 지속가능한 개발)를 포함한다. 이러한 기술들은 도시와 농지에 모두 적용되거나 혹은 기존의 다른 노하우와 혼합되어 삶의 조건을 향상시키기 위해 사용된다.

TOWARD A COMPLEX MANAGEMENT OF TERRITORIES

Our research deals with a new, biological and digital agriculture. Hydroponics, permaculture, and agroforestry hybridize with robotics to increase efficiency without polluting the soil and without prejudice to the natural balances.

This agriculture is fully automated, it is optimized and high-performance. The fields are cultivated by small robots that work on solar energy. They maintain soil, give water, mechanically weed, cut plants, cut grass, grind, pick fruits one by one, store, eliminate pests, scare birds away, put the produce in bags or boxes, and transport it. They also participate in the management and maintenance of cattle farms. Self-organized agriculture produces fruits and vegetables, biofuels and methane but also recycles urban organic waste. It purifies the water, improves biodiversity, creates landscapes, places for walking trails and gardening areas for families. Robots work night and day, silently. They maintain the trees in agroforestry and maintain an optimized period of sunshine. Robotic lombricultures produce compost, natural liquid fertilizers, and food for fish farms. The water is purified by automated phyto-ponds that produce edible plants or useful seaweeds for green fertilizers or fuels. Farms generate diversified landscapes, on the basis of bocages of small fields with mixed cultures. Urban automatic farms gather data and feed the Big Data system by transferring their information to satellites. These agricultural equipments hybridize with natural and urban environments and improve upon them. But agriculture is not only about huge machinery, it has to also produce a methodical and cultural valuing universe for mankind and it has to enable farmers to live a decent life.

The architecture of bio-digital cities is also suitable for intensive automation. The majority of effluents and waste are recycled within the buildings. Besides the high-tech agriculture, the city is designed for robots to create the link between town and countryside, also by ensuring short recycling periods. At night, the whole city is maintained with the same automated tools, just as quietly as in the fields. When the subways are not circulating, the tunnels between stations are used to sort out the goods. Wastes are compressed and placed in sealed containers. Railway wagons run on the tracks and bring fresh food directly to the various districts. They are charged and discharged by robots. Organic wastes are transported by railways into agricultural areas or composting areas. Being automated according to the principles of self-organization, the city generates an environment which is convenient for human life. Far from being a purely technical environment, the city and the architecture are based on the emergent phenomena principles. They are sensitive to the human dimension and condition. They adapt themselves in real time to the constraints, to the priorities, to the imperatives, to the needs, to the necessities of a better civilized, more tolerant, and more transparent society.

This is to take into account technologies already developed and being set up in our contemporary societies. All new techniques and new methods must be at the center of concern for sustainability. They can provide more comfort, more well-being, better living, more friendliness, tolerance, and collaboration.

Model for a high-desity
io-digital city

BAILLY / J.MAGERAND / 2013

대지의 복합관리를 지향하며

우리의 연구는 생물학적 그리고 디지털이 적용된 새로운 농업을 다룬다. 수경재배, 영속농업, 혼농임업이 로봇공학과 결합될 때 토양오염과 자연균형에 대한 침해 없이도 높은 효과를 거둘 수 있다.

이러한 농업형태는 완전히 자동화되고 고성능으로 최적화된 것이다. 경작지는 태양에너지로 움직이는 소형 로봇에 의해 일궈진다. 그들은 토양을 관리하고, 물을 공급하며, 잡초를 제거하고, 가지치기, 잔디깎기, 빻기, 과일따기, 작물의 적재, 해충의 제거, 새 쫓아내기, 생산품 포장 그리고 운송까지 해낸다. 이들은 축산농가에서도 또한 역할을 해낼 수 있다. 자가조직화된 농업은 과일과 야채를 생산하며, 바이오연료와 메탄도 만들어지는데, 이는 모두 도시 유기폐기물로 재활용된다. 물을 정화시키고, 생물다양성을 증진하며, 조경을 조성하고, 각 가정이 산책하며 정원을 가꿀 수 있는 공간을 제공한다. 로봇은 밤낮으로 조용히 일한다. 혼농임업에 있어서는, 나무를 관리하며 가장 최적화된 시간만큼 조도를 조절한다. 로봇공학은 두엄을 만들고, 천연 거름을 만들며, 양식장에 먹이를 준다. 물은 자동화된 식물연못에서 정화되는데, 이 연못은 식용가능한 작물과 거름 혹은 연료로 유용한 조류를 생산해낸다. 농장은 작은 대지와 문화가 섞여든 산림을 배경으로 다양한 풍광을 만들어 낸다. 도시의 자동화된 농장은 빅데이터 시스템으로 인공위성을 통해서 정보를 주고 받는다. 이러한 농업의 도구들은 자연과 도시 환경을 혼합시키며 동시에 향상시킨다. 그러나 농업은 단순히 거대한 기계류가 아니라 인류를 위한 체계적이며 문화적인 가치를 가진 세계를 창조하는 것이며, 또한 농업종사자가 질높은 삶을 향유하도록 하는 것이다.

바이오 디지털 시티의 농업은 또한 집중적인 자동화에 적합한 것이다. 오폐수의 대부분과 폐기물은 빌딩 내에서 재활용된다. 첨단기술의 농업 외에도, 도시는 로봇이 도심과 지방사이에 연결고리를 만들어낼 수 있도록 설계된다. 밤에는 도시 전체가 동일한 자동화 툴로 유지되는데, 마치 경작지에서 작동하는 기계와 같이 조용하다. 지하철이 운행되지 않을 때에는 역 사이사이의 터널들이 물품들을 처리하기 위해서 사용된다. 폐기물들은 압축되어 밀봉된 컨테이너들에 놓여진다. 화물열차는 트랙위를 달리며 신선한 음식을 여러 구역에 전달한다. 이 열차들은 로봇에 의해 운행된다. 유기폐기물은 철도로 옮겨져 농경지역이나 거름을 만드는 곳으로 간다. 자가 조직화(self-organizatinon)의 원리에 의하여 자동화가 이뤄지기 때문에, 도시는 인간 생활에 편리하도록 주변환경을 만들어 나간다. 순전히 기술적이기만 한 환경과는 거리가 있는 이런 도시와 건축물은 새로운 현상 원리에 기초를 둔 것이다. 이는 바로 인간 차원과 조건에 민감하게 반응하며, 스스로를 적응시켜 실시간으로 통제하고, 우선순위를 정하고, 명령하며, 수요를 파악하여 보다 나은 문화와 보다 포용도가 높고 투명한 사회로 적용해 나가는 것이다.

이는 이미 개발되고 형성된 기술을 현 사회에 받아들이는 것이다. 새로운 기술과 방법론은 지속가능성에 중점을 두고 고민되어야 좀더 안락하고 웰빙에 가까우며, 보다 나은 삶과 친목, 관용 그리고 협업으로 나아갈 수 있다.

Therefore, prototypes of digital cities are 'Hyper-cities' with a very high quality of life. They are dense, energy efficient, and save water and natural resources. They focus, expose, enhance and value sophisticated technological know-how (robotic waste treatment, smart water recycling, agricultural robotics, robotic urban storage, remote manufacturing bio-robotic restaurants, and many more). These cities contribute to re-balance in real time planetary ecosystems and agricultural lands. They include organic and high-tech urban agriculture.

Fab-labs in rural and urban areas are developed to reduce the waste of energy and raw materials. Industrial and economic organizations are re-visited by maximizing the potential of new management and information tools and by promoting short multi-recycling periods.

We believe that, more widely, all technical or industrial domains which are directly or indirectly applicable to territory, are affected by this change of production/consumption. For example, robotic agricultural equipment, environmental settlements, construction work robotics, smart remediation, automated sorting, parameterized recycling, experimental transport, computerized storage, adaptive environment, special vehicles, and smart housings, are all domains to renew industrial production methods.

This research questions all the knowledge and all of us, privately, about the mutation that we are living with today, be it cultural, civilizational, intellectual, economic, industrial or environmental. In this technical and methodical atmosphere, there is obviously the important subject of inventing new forms of collaboration between individuals, communities, societies, and countries. The traumas left by the techno-scientific utopias of the first half of the 20th century still remain. It is therefore useful, to handle with precaution, the proposals of a new environment when it claims to be inspired by new technical and scientific organizations.

The main hypothesis of this research is that the organizational dimension is transverse, that it is integral and interdependent with societal dimension. It can be enriched by IT 'substance', regardless of purely technical aspects. In this context, the question is how rational tools can inspire more cohesive, more tolerant, more humanistic groups without engaging our societies into authoritarian plans to the benefit of very few.

THE INTERNATIONAL EXPERIMENTAL STUDIO FOR BIO-DIGITAL CITY, AN EXPERIMENTAL AND HYBRID TOOL

The International Experimental Studio for Bio-Digital City has set out on a mission to develop forward-looking research and public awareness toward emerging issues related to digital. From this perspective, an important place is given to reflect on tangible or intangible solutions that could ultimately help to reduce great disorders caused by urban development and harmful agricultural practices. In addition, the Studio aims to facilitate, build and/or strengthen, technical and scientific cooperation to lead to new operational inventions and thereby generate new economic dynamics.

In August 2013, the City of Science and Industry in Paris hosted the International Studio within its prestigious exhibitions. Thus, the general public was able to encounter new approaches to the city of tomorrow. It is in this context that the Studio produces and promotes research in France, Belgium and Korea.

The Studio produces, thanks to digitals, self-organized and intelligent urban prototypes, hyper-recycling city models, real time-adaptable architecture, bio-optimized agriculture, better mutualized individual and collective organizations, that are all beneficial for the planet.

It is to identify, use, develop and cross, in urban and rural projects, the most advanced technical, scientific, industrial, sociological knowledge. From new materials, new techniques, new knowledge, and new lifestyles, the goal is to design, build and manage cities by optimizing and innovating methods (algorithms, modeling, optimization set, collaborative work etc.).

The Studio is permanent. It gathers the best young specialists versed in new tools within a flexible collaborative formula; they include architects, town planners, landscape architects, artists, engineers, computer scientists, agronomists, geographers, mathematicians, roboticists, sociologists, historians, biologists, ecologists or other experts acting on the territory of the bio-digital city.

The Studio works around collaborative platforms to work out methods of conceptual experiments incorporating integral ultra-short recycling, agricultural machinery and robotics, enhanced biodiversity, high technical and environmental housing, digital sociology etc.

따라서 이러한 디지털 시티의 프로토타입은 '하이퍼 시티', 즉 삶의 질이 굉장히 향상된 곳이다. 이곳은 밀도높고, 에너지 효율적이며, 물과 자연 자원들을 절약하는 특징을 가지고 있다. 이들은 (로봇의 폐기물 처리, 생활용수의 스마트 리사이클링, 농업용 로봇, 로봇의 도시 저장소, 원격 생산, 바이오-로봇 식당을 포함하는 수많은) 정교한 기술적 노하우에 초점을 맞추고, 노출시키며, 발전시켜나가는 동시에 가치롭게 여긴다. 이러한 도시들은 전 지구적인 관점에서의 생태계와 농업지에 실시간으로 균형을 되찾는데에 기여하는 것이다.
Fab-lab은 도시와 지방에서 에너지와 원자재의 낭비를 줄이기 위해서 개발되었다. 산업 및 경제 단체들이 새로운 관리와 정보 툴의 잠재력을 최대화함으로써, 그리고 단기 리사이클링을 다각화 함으로써 새롭게 조명되었다.
우리는 직간접적으로 대지에 적용가능한 모든 넓은 의미의 기술과 산업 영역이 생산과 소비의 변화에 영향을 받는다고 믿는다. 예를 들어, 로봇 농업 기구, 환경에 대한 합의, 건축현장에서의 로봇, 스마트 교정, 자동화된 분리기능, 매개화된 리사이클링, 실험적인 운송수단, 컴퓨터화된 저장공간, 적응제어 환경, 특별하게 개발된 운송수단들, 스마트 하우스와 같은 영역들이 산업 생산 방식을 새롭게 하고 있다.
이 연구는 모든 지식과 개별적으로 우리 모두에게 우리가 살아가고 있는 문화적, 문명적, 지식적, 경제적, 산업적 그리고 환경적으로 돌연변이인 것에 대해서 의문을 가지게 한다. 이러한 기술적이고 체계적인 분위기 속에서, 개인간, 집단간, 사회간, 그리고 나라간에 협력하는 새로운 형태를 발명하고자 하는 주요 논의가 분명히 나타나고 있다. 기술과학 유토피아의 20세기의 첫 반절 시기가 남긴 트라우마는 아직도 남아있다. 따라서 새로운 환경에 대한 제안이 새로운 과학 기술 단체에 의해 제기될 때, 매우 조심히 다루어야 유용할 것이다.
이러한 연구들의 주요한 가설은 조직적인 차원은 횡으로 가로지르는 것, 즉 사회적인 차원과 함께 필수적이면서도 상호의존적이라는 것이다. 이는 순전히 기술적인 요소와 상관없이 IT의 '실체'로서 보다 풍부해 질 수 있다. 이러한 맥락에서, 어떻게 이성적인 도구들이 (소수의 이득만을 위한 권위주위로 흐르지 않고) 보다 응집력 높고, 관용적이며, 인간적인 그룹으로 영감을 줄 수 있을지가 중요한 질문인 것이다.

국제 바이오 디지털 시티 실험 스튜디오,
실험적이고 혼합적인 도구

바이오 디지털 시티를 위한 국제 실험 스튜디오는 디지털에 관한 새로운 이슈들에 대하여 미래지향적인 연구와 대중의 인식을 발전시키기 위한 미션(mission)으로 정한 바 있다. 이러한 관점에서, 도시개발과 해로운 농업에 의해 생겨나는 문제를 궁극적으로 줄일 수 있는 유무형의 해결책에 대해 깊이 고민해볼 수 있는 중요한 장(場)이 주어진 셈이다. 게다가, 이 스튜디오는 실제 작동할 수 있는 발명들, 즉 새로운 경제역학을 만들어 낼 수 있는 과학과 기술의 협력을 돕고, 구축하며, 강화하는 것에 목표를 두고 있다.
2013년 8월 파리 국립 과학 산업관은 이 국제 스튜디오를 자신들의 명망 높은 전시에 초대했다. 따라서 일반 대중들 역시 미래의 도시에 대한 새로운 접근을 만나볼 수 있었다. 바로 이러한 맥락에서 이 스튜디오는 프랑스, 벨기에 그리고 한국에서 연구를 진행하고 후원하려 하는 것이다.
스튜디오는 디지털 덕분에, 자가조직화되며 지능을 지닌 도시 프로토타입, 고도로 재활용되는 도시모델, 실시간 변형이 가능한 건축물, 생물학적으로 최적화된 농업, 보다 상호조직화된 개인과 집단 등, 지구에 유익한 여러 개념들을 생산해 내고 있다.
이는 도시, 그리고 지방 프로젝트에서 가장 최첨단의 기술과 과학, 산업, 사회과학적 지식을 발견하고, 사용하며, 발전시키려는 것이다. 신소재, 신기술, 신지식 그리고 새로운 라이프스타일에 이르기까지, 혁신적인 방법론(알고리즘, 모델링, 최적화 세트, 공동작업 등)을 최적화 하여 도시를 설계하고, 구축하여 관리하는 것이 목적이다.
이 스튜디오는 지속적이다. 계속해서 최고의 젊은 전문가들을 모아 새로운 도구와 유연한 협업의 수식에 집어넣는다. 건축가, 도시설계자, 조경가, 아티스트, 엔지니어, 컴퓨터 사이언티스트, 농학자, 지리학자, 수학자, 로봇학자, 사회학자, 역사학자, 생물학자, 생태학자를 비롯한 수많은 전문분야가 바이오 디지털 시티라는 영역 내에서 협업한다.
이 스튜디오는 협업 플랫폼에서 개념적인 실험을 해내고 있는데, 그 주제는 초단 재활용(ultra-short recycling), 농경기기와 로봇, 생물다양성 강화, 하이테크 친환경 주거, 디지털 사회과학 등을 아우른다.

2014 BIO-DIGITAL WORKSHOP AT SEOUL CITIZEN'S HALL, AN IMPORTANT STEP IN THE RESEARCH FOR HYBRID DESIGN METHODS

The workshop organized in Seoul is part of a series of workshops organized in collaboration with the Faculty of Architecture of Yeungnam University in South Korea, EVCAU laboratory (Virtual Environments for Architectural and Urban Design), the School of Architecture of Montpellier, and the Faculty of Architecture of Mons University in Belgium. The workshops take place at the Palais de Tokyo, the Cité des Sciences in Paris, during 'Mons 2015/European Capital of Culture' event, and at Seoul Citizen's Hall.
The usual partners are the French Ministry of Foreign Affairs, Universcience in Paris, the French Institute, the French Institute in South Korea, the Association of Architects DPLG/HMONP of Korea, IUI (International Urbanism Institute), and the International Studio of Greater Paris. Research, projects and exhibitions are planned in Korea and France in 2014, 2015, 2016 (Paris, Mons, Seoul). The Studio establishes close contacts with high schools, universities, governments, high-tech industries, and local authorities.
The 2014 workshop brings together students and young professionals. They gather information and actions on teaching methods of Architecture, on the most advanced technologies, on lifestyles of the future, on energy management, on the relationship between city and climate change, and on new ways of thinking and drawing the cities of tomorrow. Participants (teachers and students) are required to reflect together on new lifestyles, new 'city modes', new agricultures, new travels, and new production/consumption, all made possible by the contemporary scientific and digital universe.

BIOGRAPHY

Within the 'International Experimental Studio for Bio-Digital City', Claire Bailly and Jean Magerand organize architectural, urban and landscape projects as well as research work and pedagogy. They are architects, landscapers, and urban planners. One holds a DEA in information science, the other holds a PhD in architecture. They have also made numerous publications and their projects have received many awards at several international landscape architecture and urban planning competitions dealing with the city of the future.

Robots for urban and agricultural works

C.BAILLY / J.MAGERAND / 2012

서울 시민회관에서 열린 2014 바이오 디지털 워크샵, 하이브리드 디자인 방법론 연구에 있어 중요한 진보

서울에서 열린 워크샵은 한국의 영남대학교 건축학부, EVCAU(Virtual Environments for Architectural and Urban Design) 실험실, 몽펠리에 건축학교 그리고 벨기에 몽 대학교 건축학부와의 협력으로 진행된 일련의 워크샵 중 한 부분이었다. 그 워크숍들은 파리 도쿄궁, 파리 과학관, 서울시시민청 그리고 '2015 몽/유럽의 문화 중심' 이벤트 동안 진행되었다.
이러한 행사는 프랑스 외무부와 파리 유니버사이언스, 프랑스 본국 문화원과 한국에 위치한 문화원, 건축 DPLG/HMONP 협회, IUI(International Urbanism Institute) 그리고 그랑파리 국제 스튜디오(International Stuido of Greater Paris)의 파트너쉽으로 이뤄졌다. 연구, 프로젝트, 그리고 전시가 한국과 프랑스에서 2014, 2015 그리고 2016년(파리, 몽, 서울)에 각기 계획되었으며, 고등학교, 대학교, 정부, 첨단기술 산업 그리고 각 지역기관들과 긴밀한 관계를 발전시켜왔다.
2014년 워크샵은 학생들과 젊은 전문가들을 한곳에 단합시켰다. 그들은 건축과 최신 기술, 미래의 라이프 스타일, 에너지 관리, 도시와 기후 변화, 그리고 도시의 내일을 고안하고 그려내는 새로운 방법에 대하여 정보를 모으고 실천방안을 고민했다. 참가자들(튜터들과 학생들)은 현 과학과 디지털 세계로 가능하게 된 새로운 라이프 스타일, 새로운 '도시생활양식', 새로운 건축, 새로운 여행, 새로운 생산과 소비에 대해 함께 고민해 보았다.

'바이오 디지털 시티를 위한 국제 실험 스튜디오'(International Experimental Studio for Bio-Digital City)에서, 끌레어 바이이(Claire Bailly)와 장 마젤랑(Jean Magerand)은 건축, 도시, 그리고 조경 프로젝트를 연구와 교육과 더불어 기획했다. 그들은 건축가, 조경가이자 도시설계자이다. 끌레어 바이이는 정보과학에서 학위(DEA)를, 장 마젤랑은 건축 박사학위(PhD)를 보유하고 있다. 이들은 수많은 출판물을 발행했으며, 이들의 프로젝트는 미래의 도시를 주제로 한 몇몇 국제 조경 건축 그리고 도시계획 대회에서 수상했다.

Marlène VERNOUILLET, Scientific facilitator
At the Cité des sciences et de l'industrie, Universcience, Paris

THE ROLE OF FACILITATORS IN WORKSHOP PROGRAMMES

FOREWORD

I took part in the workshop in Seoul from the 5th to the 14th of August, 2014, which followed from a series of workshops organised at the Cité des sciences in 2013. For the Cité des sciences, this workshop formed part of a broader program of events designed to include the general public in discussions about the urban habitat and rural environments.

AN URBAN MODEL AT BREAKING POINT

There is an urgent need for all of us - architects, urban planners, designers and residents - to rethink the city of the future in a way that protects agricultural land. It will be vital to 'reinvent living spaces' in the years ahead because, in numerous ways, our urban models are increasingly flawed.
As the global population increases, so does the number of people living in our towns and cities. Since 2007, more people now live in urban areas than in the countryside. By 2030, an estimated two thirds of the world's population will be living in urban areas.
The world's population is also getting older. More than 606 million people are currently aged over 60, 379 million people are aged over 80. By 2050, the number of people aged over 60 and 80 will increase fivefold and threefold respectively. What changes do we need to make to the urban environment to meet these people's specific needs, especially with respect to their poor mobility?
Social inequality has also increased, and gentrification - when wealthier residents occupy a space initially inhabited by less advantaged occupants or users, transforming the economic and social make-up of the neighbourhood to the exclusive benefit of higher social strata - is gaining ground. Having little contact with their neighbours, people are increasingly isolated and few standard urban residential areas are equipped with spaces where people can meet and be together.
From an environmental point of view, the world's resources are declining while pollution and domestic energy costs continue to climb.
Lastly, the quality of the urban living environment has also taken a turn for the worse. There is a shortage of affordable housing in urban areas and the number of shanty towns is increasing.
At a time when the urban population is growing, we need to find ways to build living spaces that reflect everyone's desires and needs, while preserving the planet and its resources. Is it possible to strike a balance between farmland and urban areas? There is nothing new about these questions and people are already beginning to develop small-scale solutions to these problems. Eco-neighbourhoods, increasingly popular in Europe, are just one example. Others are taking practical steps to develop new building materials or to design residences with community areas. People are also coming together to reflect on these issues, and these workshops form part of this effort.

The workshop at the Cité des sciences / Summer 2013

서문

2013년 파리 과학관(Cité des sciences) 행사에 이어 2014년 8월 5일부터 14일까지 서울에서 개최된 워크샵에 참여하였다. 파리 국립과학관에 있어서 이번 워크숍은 일반 대중이 도시 거주지와 지역 환경에 대한 논의에 참여할 수 있도록 하는 프로그램의 일환이다.

한계점에서의 도시모델

건축가, 도시계획자, 디자이너 그리고 거주자를 아우르는 우리 모두에게 농업지를 보호할 수 있는 방향으로 도시의 미래를 재고하는 것이 긴급히 필요하다. 이는 우리의 도시모델이 점점 더 많은 결함을 나타내고 있기 때문에, 향후 '삶의 공간을 재창조'함에 있어서 여러 가지 이유에서 필수적인 요건이 될 것이다.
전세계 인구가 점점 증가함에 따라, 마을과 도시에 거주하는 인구도 점점 늘어나고 있다. 2007년부터 시외곽에 거주하는 사람들보다 도시에 거주하는 사람들이 늘어났다. 2030년에 이르면, 세계 인구의 3분의 2가 도시지역에서 거주할 것으로 예상된다.
또한, 세계인구는 점점 고령화로 가고 있다. 6천6십만명 이상이 사람들이 현재 60세 이상이며, 3천7백9십만명은 80세 이상이다. 2050년에 이르면, 60세 이상의 인구는 5배, 80세 이상의 인구는 5배 증가할 것이다. 이러한 인구(이들 대부분이 거동이 불편함을 고려했을 때)의 특정한 수요를 만족시키기 위해서는 우리가 도시환경에 어떤 변화를 만들어 내야 할까?
사회의 불평등 또한 증가해왔다. 고급주택화(젠트리피케이션), 즉, 보다 부유한 거주자들이 상대적으로 빈곤한 기존 거주자들의 공간을 차지해 버리고, 그 지역의 사회 경제적 구성을 상위계층에게 유리하도록 변형시켜 버리는 현상이 더욱 강력해지고 있다. 이웃들과의 교류나 접촉이 적어짐으로써, 사람들은 점점 고립되어 가고, 전형적인 도시거주지역들 중 사람들이 서로 만나고 함께 어울릴 수 있는 공간을 갖추고 있는 곳은 거의 없다. 환경적인 관점에서, 환경오염과 국내 에너지 비용이 점점 증가하는 것에 반해, 전 세계의 자원은 점점 감소하고 있다.
마지막으로, 도시의 주거 환경이 점점 더 악화되고 있다. 사람들이 구입할 수 있는 가격의 집은 도시에서 점점 사라지고, 판자촌이 증가하고 있는 추세이다.
도시인구가 증가하고 있는 시점에서, 우리는 모든 이들의 욕구와 필요를 반영하는 삶의 공간을 건축할 수 있는 방법을 찾아야 하며, 이는 지구와 자원을 보존하는 방식이어야 한다. 과연 도시공간과 농업지의 균형을 만들어 내는 것이 가능할 것인가? 이러한 질문은 더 이상 새로운 것이 아니다. 사람들은 이미 작은 규모로 해답을 만들어 나가고 있기 때문이다. 그 한 예인 생태 주민공간(Eco-neighbourhoods)은 유럽에서 점점 더 인기를 얻어가고 있다. 새로운 건축재료를 개발하는 것, 혹은 공동 공간을 가진 거주지역을 설계하는 등의 다른 대안도 실질적인 진전을 이루고 있다. 사람들은 점점 더 이러한 이슈들을 고민하기 위하여 함께 모이고 있으며 이러한 워크샵들이 바로 그 노력의 일환이다.

Cité des sciences workshop poster / 2013

Seoul Citizen's Hall workshop poster / 2014

THE 2013 WORKSHOPS:
USING FACILITATORS TO INVOLVE THE GENERAL PUBLIC IN AN EXPERIMENTAL LABORATORY

We have organized several activities in partnership with the Atelier international expérimental pour la Cité Bio-numérique, including 'The digital city and urban, organic, high-tech agriculture': an international event consisting of a consecutive series of five workshops held between the 13th of July and the 31st of August, 2013. Each lasted ten days and explored one of the following themes: parametric processes, automatic control, pooling, hybridisation and self-organisation. The workshops were moderated by the architects, urban planners and landscape designers Claire Bailly and Jean Magerand.

Some 99 young professionals and students from Korea, Brazil, Belgium, Italy and France were asked to work on a specific foresight issue: how to develop urban areas without encroaching on agricultural land.

The participants were asked to imagine a high-density, highly-adaptable, highly-sustainable city, while integrating productive, organic and high-tech agriculture into the urban environment by using or imagining leading-edge technology.

This workshop was held at the heart of the Cité des sciences, with the intention of opening up the discussions to the general public. The role of the facilitators was to place the general public at the centre of the foresight activities, foster debate on the city of the future and the development and planning of sustainable urban environments, provide an insight into current research, disseminate and break down the barriers between knowledge, and finally, involve young professionals and students in the debate. To put these goals into practice, we created four spaces:

- The first was a mini-conference area where experts gave daily talks on their specialism within their professional field and shared their experiences to inform the debate.
- A work area for participants. Facilitators organised two visits a day to the area to enable these participants to meet and talk with the general public. These events fostered an essential exchange of ideas between residents and 'future builders'.
- A display area for models produced in the on-site Fab Lab: a practical tool that allowed people to illustrate ideas discussed within a facilitation context.
- A digital area equipped with tablets where the public were invited to imagine and design the homes of the future.

These activities enabled more than 22,000 visitors to understand the challenges involved in developing the cities of the future, discover current thinking on the issue and express their ideas through drawing. Workshop participants were also able to talk with the public and gain an insight into their doubts and expectations. The public's involvement in these foresight activities was therefore mutually beneficial, challenged prejudices and changed the way people see urban models.

In parallel to these workshops, the facilitation team organised a programme of activities on the same theme, including scientific workshops on eco-neighbourhoods, grey energy and the factors that shape the urban environment.

The workshop at the Cité des sciences / Summer 2013

2013년 워크샵:
일반 대중을 실험연구소로 끌어들이기 위한 협력자의 활용

우리는 Atelier international expérimental pour la Cité Bio-numérique과 파트너쉽을 맺고 몇몇 활동을 진행해왔다. 2013년 7월 13일부터 8월 31일까지 개최된 '디지털 시티와 도시, 유기체, 그리고 첨담기술의 농업'(The digital city and urban, organic, high-tech agriculture), 즉 5개의 워크샵이 연속으로 열린 국제 행사 역시 그 중 하나였다. 각각의 워크샵은 열흘간 계속되며 파라메트릭 프로세스(Parametric processes), 자동조절장치(Automatic control), 풀링(Pooling), 교배(Hybridisation), 자기조직화(Self organisation)와 같은 주제 중에서 하나를 선택하여 탐구하였다. 워크샵들은 건축가, 도시설계자, 조경가인 끌레어 바이이(Claire Bailly)와 장 마젤랑(Jean Magerand)에 의해 중재되었다.

한국, 브라질, 벨기에, 이탈리아 그리고 프랑스에서 모여든 99여명의 젊은 전문가들과 학생들이 예견된 이슈들에 대해 함께 협력하도록 요청받았다. 바로 "어떻게 농지를 침해하지 않고 도시를 개발할 수 있을까"라는 주제였다.

참가자들은 높은 밀도와 적응력이 좋고 높은 지속가능성을 가지는 도시를 상상하도록 요구받았는데, 생산적이고 유기적이면서 첨단기술을 갖춘 농업을 도시지역으로 끌어들이는 것이 그 과제였다.

이 워크샵은 파리 과학관(Cité des sciences)의 중앙홀에서 개최되었고, 이는 일반 대중들에게 논의를 개방하고자 하는 의도였다. 협력자(facilitator)들의 역할은 일반 대중이 예견된 활동들에 참여할 수 있도록 하며, 도시의 미래와 지속가능한 도시 환경의 개발에 대하여 논의할 수 있도록 하고, 현재 진행되고 있는 연구와 지식을 제공하여 젊은 전문가들과 학생들이 활발히 논의에 참여하도록 하는 것이다. 이러한 목적을 달성하기 위해서 우리는 네가지 서로 다른 공간을 만들었다:

- 첫 번째는 세미나를 할 수 있는 공간으로, 각 전문가들이 해당 분야에 대한 지식과 경험을 토론에 매일 제공하는 형태이다.
- 두 번째는 참가자들을 위한 작업 공간. 협력자들은 일반 대중과 참가자들이 만나고 대화를 나눌 수 있도록 이 작업 공간에 하루에 두 번 방문을 주도한다. 이 이벤트는 '미래 건설자들'과 거주민간의 교류와 아이디어 공유를 이끌어 낸다.
- 세 번째는 Fab Lab, 즉 사람들이 협력자들의 도움으로 논의된 아이디어들을 표현할 수 있도록 돕는 현실적인 도구에서 만들어진 모형을 전시하는 공간.
- 그리고 일반 대중이 초대되어 미래의 주택을 상상하고 디자인해 볼 수 있도록 타블렛이 구비된 디지털 공간.

이러한 활동들은 2만 2천명 이상의 방문객들이 미래 도시를 개발하는 데에 연관되는 어려움을 이해하고, 현재 이 이슈들과 관련된 연구를 이해함은 물론, 자신들의 생각을 그림을 통해 표현할 수 있도록 해주었다. 워크샵 참가자들 역시 대중과 소통하며 그들의 생각과 기대를 들을 수 있었다. 이러한 활동에서 대중의 참여는 상호간 이득을 줌은 물론, 선입견을 불식시키며, 사람들이 도시모델에 대해 생각하는 방식을 바꾸어 놓는다.

이 워크샵들과 동시에, 협력자 팀들은 동일함 테마의 프로그램을 구성하였는데, 이는 생태 주민공간(eco-neighbourhoods), 회색 에너지, 도시환경의 요소와 형태에 대한 과학 워크샵과 같은 프로그램을 포함하는 것이었다.

INTERNATIONAL WORKSHOPS PAST AND FUTURE

The Seoul international workshop (media and architecture) in August 2014, 'Bio digital city: tomorrow's Seoul project', was an extension of these workshops and took a foresight approach to the city.
The presence of scientific mediation of the Cité des sciences has been possible thanks to the collaboration with the French Institute.
The workshop took place in the Citizens' Hall, part of Seoul City Hall. A lot of people walk through the area or use it as a central meeting point. The choice of the site was therefore symbolic, since the city's residents will play a vital role in thinking about the city of the future. The international professionals who coordinated the workshop benefited participants with the feedback and input of their peers, and introduced a cultural dimension into the proceedings. This was especially the case in Seoul, since the participants were also of different nationalities.
Works of art were exhibited close to the workshop venues, attracting more visitors and giving them a chance to consider these issues from a different point of view.
These workshops have grown in strength and complexity over time, allowing us to make advances, both in terms of ideas and practices.
Building on this workshop experience, we are currently planning to organize similar activities as part of the Year of Korea in France at the Cité des sciences in 2015, then in Mons (Belgium), European Capital of Culture, in February 2015, and lastly to coincide with the Year of France in Korea in August 2016.

Universcience is the first French public establishment for the dissemination of scientific and technical culture born out of the merger in 2010 between the Cité des sciences et de l'industrie and the Palais de la découverte. Its 3 million visitors per year have made Universcience the 4th most visited cultural site in France and the leading science centre in Europe. Its interactive and fun approach is aimed at tackling science and techniques through the discovery of the major challenges facing our world. Its offers range from temporary or permanent exhibitions, spaces dedicated to children, resource facilities, experimentation with innovative digital tools.

Universcience shares its know-how in France and internationally with a view to spreading scientific and technical culture.

- touring exhibitions, from 50 sq.m to 700 sq.m, for all audiences and covering a wide range of topics;
- educational spaces dedicated to children, "Cité des enfants";
- cultural engineering and museum consulting.

과거와 현재의 국제 워크샵

2014년 8월에 개최된 서울 국제 워크샵(미디어와 건축), '바이오 디지털 시티: 내일의 서울 프로젝트'(Bio digital city: tomorrow's Seoul proejct)은 이러한 워크샵들의 연장선이며, 또한 미래를 내다보고자 하는 노력이다.
파리 과학관의 과학적인 중재노력은 프랑스문화원의 협력 덕분에 가능하였다.
이 워크샵은 서울시청의 시민청에서 이뤄졌는데, 많은 사람들이 지나다닐 뿐만 아니라 미팅장소로 활용하는 곳이다. 이러한 장소의 선택은 도시의 미래를 고려함에 있어서 거주 시민들이 매우 필수적인 역할을 할 것임을 상징하는 것이다. 워크샵을 조직한 국제 전문가들은 참가자들이 피드백과 동료들로부터의 조언을 충분히 받을 수 있도록 돕고, 그 과정에서 문화적인 차원을 도입하였다. 이는 특히 서울에서 드러난 모습이었는데, 참가자들이 매우 다양한 국적을 가지고 있었기 때문이다.
예술 작품들은 워크샵 장소 근처에 전시되었으며, 보다 많은 방문객들을 끌어들여 그들이 각기 다른 관점에서 이슈를 생각해 볼 수 있는 기회를 제공하였다.
이러한 워크샵들은 시간이 지남에 따라 점점 더 강력해 지고 복잡성을 더해가, 우리에게 새로운 구상과 실현으로 발전해 나갈 수 있도록 한다.
워크샵에서의 경험을 바탕으로, 현재 우리는 '프랑스에서 한국의 해' 행사 중 하나로, 2015년에 국립 과학관에서 유사한 행사를 기획하고 있다. 그리고 유럽 문화의 중심인 벨기에 몽에서 2015년 2월에, 그리고 2016년 8월에는 '프랑스 한국의 해' 행사와 동시에 또 개최될 예정이다.

cité des sciences & de l'industrie · Palais DÉCOUVERTE · by universcience Paris

KIM Kihong
3D Print

ARCHITECTURE MEET ART

REVERSE-ENGINEERING

ENCOUNTER OF ARCHITECTURE AND ART

This piece of work is a sequence of interesting work that creates a work through cooperation of an architect and an artist. It tried to add the properties and depth of materials by adding digital techniques to the drawing and fiber molding of artist Oh Hwajin. First of all, it required digital data work of the fiber molding of artist Oh Hwajin through 3D scanning. This course went through a work to develop a scanned dot file to a face and this repeated several face revision stages and formed a soft curve. Then, it reached a final stage by applying the pattern designed by an architect to a shape.

FORM

The trophy work for the bio-digital exhibition is a process to impromptu and random cut artificial leather and paste and assembly it impromptu. This working method was not attempted for the first time but stemmed from the work of "What's the figure?" that was progressed by me.

RELAXATION

WORKING PROCESS

After composing the dot file databased through 3D scanning to a rough face, I tried to maintain it as the original shape as much as possible, optimizing it to the best stable shape. This is the most important work as a preceding stage to apply the properties at the next stage. This course is the most difficult work and one of complicated stages. Shaping using not a surface but a mesh goes through pushing and pulling one object in a designed shape to make a form and the soft catmull clark subdivision method. After this course, it completes the stage of mesh relaxation and the course for point stabilization of the mesh face. At this point, the main work is to use the kangaroo and weaver bird, etc. of grasshoppers with the algorithm tool. Then, the design shape of a pattern planned by me can be covered. This can highlight the feeling of properties and the depth of geometry by using patterns.

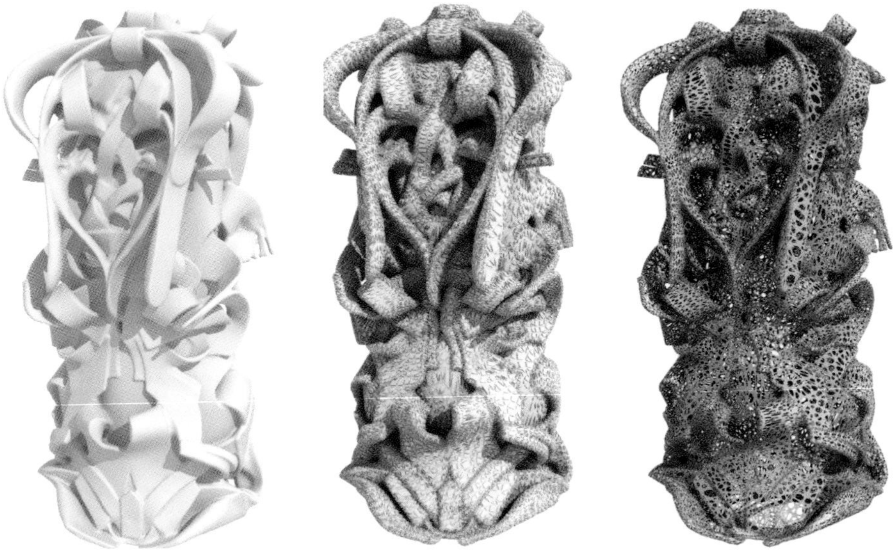

분해공학

건축과 예술의 만남

이번 작품은 건축가와 예술가가 협력하여 하나의 작품을 만들어 내는 흥미로운 작업의 연속이었다. 오화진 작가의 드로잉과 섬유소조에 디지털 기법을 더하여 재료의 물성과 깊이를 더하고자 하였다. 먼저 오화진 작가의 섬유소조물을 3d scanning을 통해 디지털 데이터화 시키는 작업이 필요 했다. 이 과정에서 스캔받은 점파일을 면으로 발전시키는 작업을 진행하였으며, 이는 몇차례의 면 수정 단계를 거쳐 부드러운 곡면을 형성하게 되었다. 여기에 건축가가 디자인한 패턴을 형상에 입힘으로써 최종 단계에 이르게 된다.

형태

바이오 디지털 전시를 위한 트로피 작업은 인조 세무를 즉흥적이고 무작위적으로 잘라내어 이것을 다시 즉흥적으로 붙이고 조립하여 만든 작업이다. 이러한 작업방식은 이번에 처음 시도한 것이 아니라 기존에 본인이 진행하던 "What's the figure?" 라는 작업에서 비롯한 것이다.

안정화

진행과정

3d scanning을 통해 데이터화한 점파일을 거친면으로 구성한 다음, 이를 가장 안정적인 형태로 최적화 해가며 원래의 형태를 최대한 유지하고자 하였다. 이는 다음단계에 있을 물성을 적용하는 과정의 선행 단계로서 가장 중요한 작업이다. 이 과정이 가장 어려운 작업이며 복잡한 단계중 하나이다. Surface가 아닌 mesh를 이용한 형태 잡기는 하나의 오브젝트를 계획된 형상에 맞게 밀고 당겨서 형태를 잡고 catmull clark subdividion 방식으로 부드럽게 한다. 이과정이 끝나면, mesh relaxation 단계를 거쳐서 mesh면의 point 안정화 과정이 이어진다. 이때 알고리즘 tool로 grasshopper의 kangaroo/weaver bird 등을 이용한 작업을 주로하게 된다. 이렇게 면작업이 끝나면, 내가 원하는 패턴의 디자인을 형상에 입힐 수 있다. 이는 물성의 느낌을 살릴수도 있으며, 패턴을 이용하여 geometry의 깊이감을 살릴 수 있다.

< ALGORITHM CODE _Computer lanuage >

```
/// <summary>
/// This class will be instantiated on demand by the Script component.
/// </summary>
public class Script_Instance : GH_ScriptInstance
{ #
region Utility functions
/// <summary>Print a String to the [Out] Parameter of the Script component.</summary>
/// <param name="text">String to print.</param>
private void Print(string text) { /* Implementation hidden. */ }
/// <summary>Print a formatted String to the [Out] Parameter of the Script component.</summary>
/// <param name="format">String format.</param>
/// <param name="args">Formatting parameters.</param>
private void Print(string format, params object[] args) { /* Implementation hidden. */ }
/// <summary>Print useful information about an object instance to the [Out] Parameter of the Script component. </
summary>
/// <param name="obj">Object instance to parse.</param>
private void Reflect(object obj) { /* Implementation hidden. */ }
/// <summary>Print the signatures of all the overloads of a specific method to the [Out] Parameter of the Script
component.
</summary>
/// <param name="obj">Object instance to parse.</param>
private void Reflflect(object obj, string method_name) { /* Implementation hidden. */ }
#endregion
#region Members
/// <summary>Gets the current Rhino document.</summary>
private readonly RhinoDoc RhinoDocument;
/// <summary>Gets the Grasshopper document that owns this script.</summary>
private readonly GH_Document GrasshopperDocument;
/// <summary>Gets the Grasshopper script component that owns this script.</summary>
private readonly IGH_Component Component;
/// <summary>
/// Gets the current iteration count. The first call to RunScript() is associated with Iteration==0.
/// Any subsequent call within the same solution will increment the Iteration count.
/// </summary>
private readonly int Iteration;
#endregion
/// <summary>
/// This procedure contains the user code. Input parameters are provided as regular arguments,
/// Output parameters as ref arguments. You don't have to assign output parameters,
/// they will have a default value.
/// </summary>
private void RunScript(Mesh M, double P, ref object A)
{
if(M != null){
Guid meshObj = doc.Objects.AddMesh(M);
doc.Objects.UnselectAll();
doc.Objects.Select(meshObj);
Rhino.RhinoApp.RunScript("_-ReduceMesh _ReductionPercentage " + Convert.ToString(P) + " _Enter", false);
Rhino.DocObjects.RhinoObject MObj = doc.Objects.GetSelectedObjects(false, false).First();
doc.Objects.Delete(meshObj, true);
A = MObj.Geometry;
}
}
// <Custom additional code>
// </Custom additional code>
```

POINT CLOUD

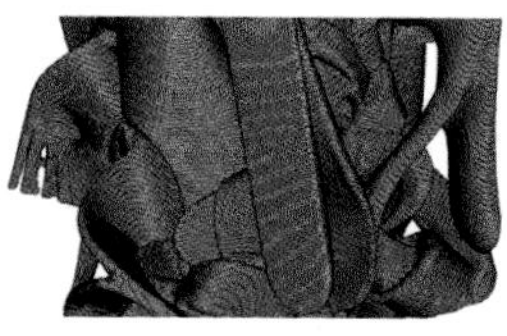

MESH DEVELOPE

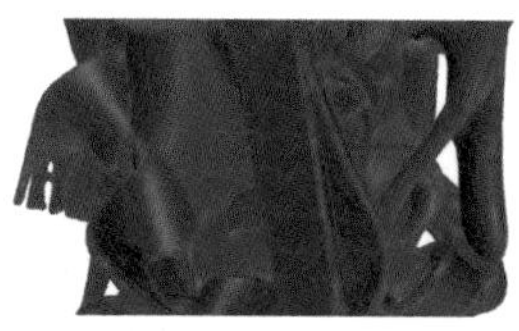

MESH RELAXATION

ALGORITHM DISIGN

PATTERN

When seeing the work of artist Oh Hwajin, I strongly felt it weird and grotesque vaguely. Then, I was captured by the thought to put some patterns or properties that can emphasize this feeling. At first, I attempted circle packing, voronoi, irregular and attractor pattern, but I felt that this rather contrasts the original shape to reduce the characteristics of the existing shape. Therefore, I got to have a plan to apply engraving and embossing to express the irregular of geometry with a complicated shape as it is. It could keep the unique feeling of the original shape by expressing a simple pattern by engraving and embossing below the same order of rank.

ENCOUNTER OF ANALOGUE AND DIGITAL

I sincerely give a clap to the enthusiasm of artist Oh Hwajin who worked with interest in the new working method that was different from his formal methods. Even though I was excited and worried about the progress to cooperate each other and database in the analogue work from the digital method, but it was actually an interesting progress very much. I am satisfied very much with the result and process through the encounter course of using fabric modeling and computation,

알고리즘 디자인

패턴

오화진 작가의 작품을 보았을때 처음 느낀 감정은 막연히 괴상하고 그로테스크하다는 느낌을 강하게 받았다. 그렇다면 이런 느낌을 강조할 수 있는 패턴 혹은 물성을 입혀 보자는 생각이 강하게 자리 잡았다. 처음에는 circle packing, voronoi, irregular/attractor pattern을 시도해 보았으나, 이는 원형상과 대조되어 오히려 기존 형상의 특징이 반감되는 듯한 느낌을 많이 받았다. 그래서 생각한 것이 복잡한 형태의 geomerty가 가지는 불규칙한 조직의 느낌을 그대로 음각, 양각을 적용하여 표현하고자 하였다. Geometry의 똑같은 위계질서 아래 단순히 패턴을 음각과 양각으로 표현함으로써 원형상이 가지는 독특한 느낌을 그대로 살릴 수 있게 되었다.

아날로그와 디지털의 만남

기존 자신이 고수해 오던 방식과 다른 새로운 형태의 작업방식에 흥미를 가지고 진행해 주신 오화진 작가님의 열정에 박수를 보낸다. 아날로그 작업에 디지털 방식으로 데이터화 해가며 서로 호흡을 맞출수 있는 진행방식에 기대와 걱정이 교차하였지만, 사실 매우 흥미로운 진행과정이었다. 섬유소조(fabric modeling)와 computation을 이용한 작업이 조우하는 과정을 통해 오늘의 결과물과 과정을 만든 것에 매우 만족한다.

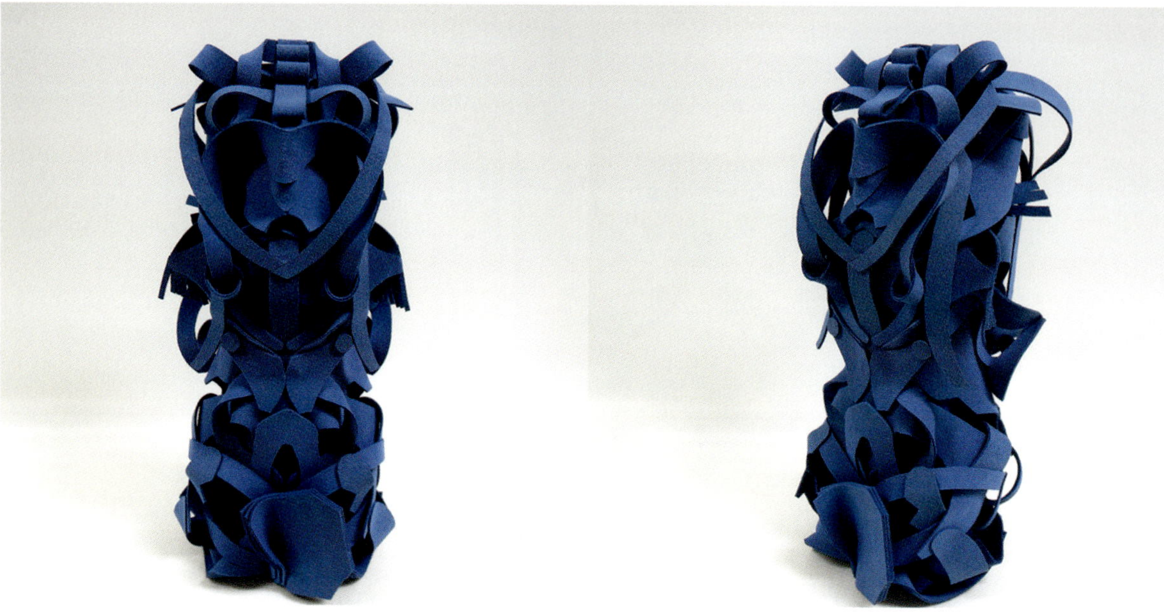

REDUCED MESH

SUB-DIVISION MESH

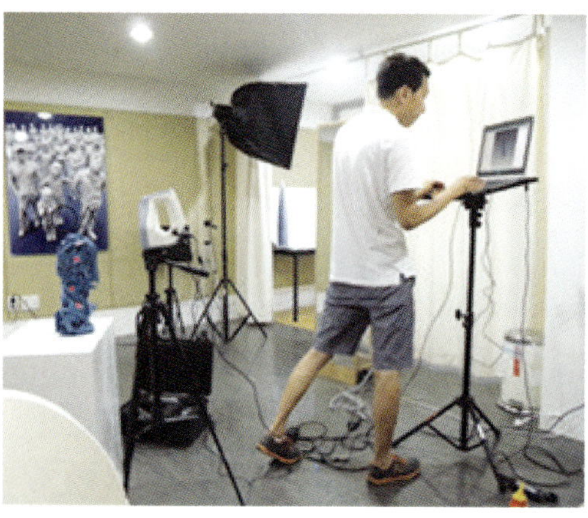

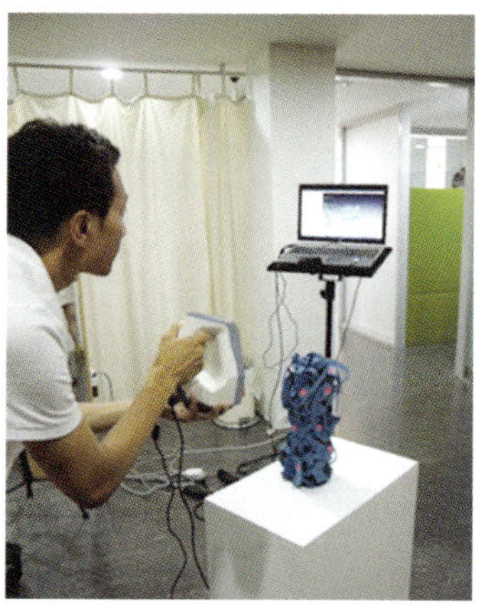

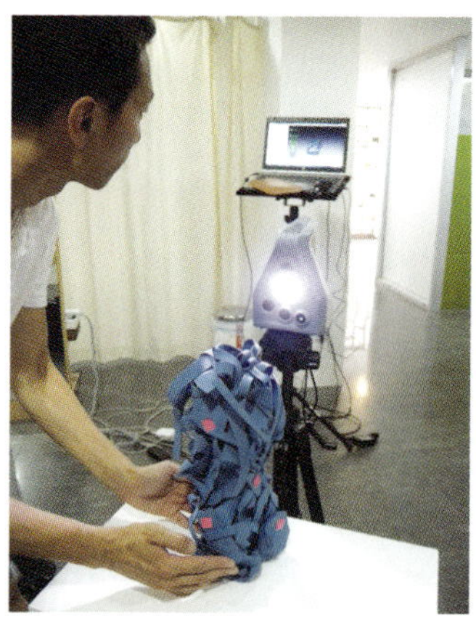

architercture
landscape
media

T - A

KIM Guyhong

Designing the future value of urban agriculture in Seoul, a 600 year-old capital city, is not a simple task. We had to recollect and link many components of Seoul in order to propose a new paradigm for urban Seoul. Moreover, results produced during the "Miracle on the Han River" – an era of rapid growth in Korea after winning its independence – were also vestiges of Seoul that could not be neglected. As such, we decided to propose the value of a future Seoul on the peripheries of the Han River.

Seoul can be divided into Gangbuk – the old urban center – and Gangnam – a district developed during the industrial era by the Han River, a river which brings history and scenery together, at its center. For Seoulites, the Han River is a place to rest and a natural axis that divides the city's identity. Also, it is a panorama slider that offers a compressed recollection of Seoul's growth from the Joseon dynasty, onto the colonial period, modern development era, and the present.

We decided to investigate a new version of urban agriculture with riverside apartments near the Han River constructed during the industrial age. This was based on our desire to begin with an inquiry into possible lifestyles for Seoulites residing in apartment complexes located near the Han River, a symbolic location in Seoul. Ultimately, our design will be a proposal regarding the residential environment of Seoulites that allows future residents to create a more ecological and human-friendly urban Seoul.

Established during the early stages of Gangnam's development, the Apgujeong district is at the center of Gangnam's residential zone and boasts an excellent view of the Han River. However, the exclusive nature of this residential area prevents access from nearby areas to the riverside, and the area is closed off from the surrounding neighborhoods; therefore, we must seek solutions that can find ways to let the urban residential block coexist with surrounding areas from an urban perspective.

The Ichon area is located on the north-south axis greenbelt, formed along the mountains in Seoul, and it is the geographical center of the city that faces the Han River. It is a site in need of interpretation with regards to its sense of place at the center of the Han River development, which is at a meeting point between the north-south axis that connects Bukhansan Mountain, Namsan Mountain, Yongsan-gu, the Han River, and Gwanaksan Mountain and the east-west axis that encompasses the Han River development zone.

600년의 수도 서울에서 도시농업이 가지는 미래의 가치를 디자인한다는 것은 그리 간단한 작업은 아니다. 서울에 새로운 패러다임을 제안하기 위해, 우리는 서울의 많은 부분들을 기억해내고 연결해야 했다. 게다가 해방 이후 한강의 기적으로 불리는 급성장 시기의 산물 또한 놓쳐서는 안되는 서울의 흔적 중 하나이다.

그렇게 우리는 한강 주변에서 미래서울의 가치를 제안하기로 했다.

서울은 역사와 경관을 아우르는 한강을 기준으로, 구 도심지역인 강북과 산업시기에 개발된 강남지역으로 나뉘어진다. 한강은 서울시민들의 휴식의 공간이자 도시의 정체성을 가르는 자연적인 축이다. 또한 조선시대, 식민시대, 근대개발시기, 현대의 자취들이 어우러져 도시의 성장을 압축적으로 보여주는 파노라마 슬라이더이기도 하다.

우리는 여기 산업시대에 건설된 한강변 아파트단지들을 대상으로 도시농업의 새로운 버전을 탐구해 보기로 했다. 이는 한강이라는 상징적 장소에 위치한 공동주거단지에서 서울의 시민들이 어떤 라이프스타일을 가질 수 있는지에 대한 탐구로 시작하고 싶어서이다. 결국 우리가 디자인하는 것은 미래 서울 시민들이 더 생태적이고 인간적인 서울을 만들어갈 수 있는 시민들의 생활환경에 관한 하나의 제안이 될 것이기 때문이다.

압구정 지구는 강남개발 초기단계에 형성된 한강변 조망권이 우수한 강남주거의 중심지역이다. 하지만 주거단지의 폐쇄성으로 인근지역에서 강변으로의 접근을 막고, 주변지역과도 단절되어있어, 어떻게 하면 도시적 차원에서 도심 주거 블록이 주변지역과 상호존중을 할 수 있는가에 대한 해답을 고민해야 한다.

이촌 지구는 서울의 남북 산악녹지축 선상에 위치하며 한강과 마주하고 있는 서울의 지리적 중심지역이다. 북한산, 남산, 용산, 한강, 그리고 관악산으로 이어지는 서울의 남북축 선상에, 동서를 잇는 한강변 개발지역의 중심에 위치하는 장소성에 대한 해석이 필요한 장소이다.

LEE Jongeun

Since industrialization, vast areas of farmland producing food has been expelled from the downtown area. This is owed to the constant increase in the price of land. The project of the A-1 team is meaningful in terms of studying a spatial way to return the loss of agricultural space. Urbanites lost their agricultural space due to their desire to gain capital and it has had a negative effect on their daily life. There is a suggestion of diverse city agricultural space that is split into small units that could be suitable for city dwellers' lifestyles: vegetable gardens on balconies where you can harvest fresh lettuce, scallions, and so on for your dinner table, core farms inside elevator halls where you can communicate within your residential neighborhoods, as well as markets on conveyer belts in which daily cultivated vegetables rotate inside the complex. Those elements would be linked by another strong element, what we call a BIO BELT circulating through the complex and unifying the community as a whole.

The project of the A-2 team originates from a critical mind about the BIO-DIGITAL CITY in two ways. The first concern would be how to implement a new city agricultural program among existing cities from the point-of-view of urbanism. They prefer to choose environmentally friendly solutions which acknowledges existing public order and finds adjustable methods, rather than to eliminate past memories and set a new order. The second concern is about a methodological matter: how to have urbanites, who are busy with daily life, meet the demands of an agricultural program with much needed sincerity and effort. There is a suggestion about the combination of an automated robot cultivation system that links all of the complex together and smart applications. With this suggestion, city people can experience the process of life circulation in their daily life with no difficulty.

Team A-3 runs the project by focusing on the study of the system either about how agriculture gets into city organization, forms a connection with other programs like residences, businesses and commercial facilities or how agriculture transforms according to the changes of time and situations. The jungle gym-shaped 3D grid system which team A-3 proposes functions as a platform where individual program modules could be inserted or ejected as occasion demands like Jenga. Along the 3D grid system, there is a formulation of green networks in which simple crops could grow. Full-scale of space for city agriculture is suggested as a module of Jenga.

산업화 이후, 지속적인 지가상승과 함께 우리의 먹거리를 책임지던 드넓은 논밭들은 점점 더 도시 외곽으로 밀려났다. A-1팀의 프로젝트는 자본의 논리에 의해 도시인이 잃어버린 농업공간을, 잠들고 일어나고 쇼핑하고 이웃과 함께하는 그들의 소소한 일상 속으로 돌려주는 공간적 방법의 탐구라는 점에서 의미가 있다. 저녁식탁에 바로 뽑은 신선한 상추, 파 등을 제공할 발코니 텃밭, 이웃간 소통의 주제가 될 수 있는 엘리베이터홀 내 core farm, 카페 레스토랑에 숨쉬는 인테리어 요소로 작용할 vertical garden, 그날 그날 생산된 농산물들이 단지 전체를 순환하는 컨베이어벨트 위에서 판매되는 마켓 등, 도시인의 라이프사이클에 맞게 작은 단위로 분절된 다양한 도시농업공간이 제안되며, 이러한 요소들은 단지 전체를 순환하는 Bio Belt라는 강력한 요소에 의해 연결되며 하나로 완성된다.

A-2팀의 프로젝트는 Bio Digital City라는 주제를 두 가지 문제의식으로부터 접근한다. 첫째는 도시계획적 관점으로서 도시농업이라는 새로운 도시프로그램을 기존의 도시조직 속에서 어떠한 방식으로 실행할 것인가 하는 문제이다. 이들은 기존의 기억을 일소하고 새로운 질서를 부여하는 tabla rasa가 아니라, 존재하는 질서를 인정하고 그 위에서 조정 가능한 해법을 찾는 좀 더 자연친화적인 방식을 선택한다. 둘째는 많은 정성과 수고로움이 필요한 농업이라는 프로그램과 바쁘고 여유 없는 생활 속의 도시인이 어떻게 만날 수 있도록 하는가 하는 방법론적 문제이다. 단지전체를 연결하는 선형의 자동화된 로봇 재배시스템과 스마트 어플리케이션의 결합이 그 해법으로 제안되어 도시인이 커다란 어려움 없이 자신의 일상 속에서 생명순환의 과정을 경험할 수 있도록 하였다.

A-3팀은 농업이 주거, 업무시설 혹은 상업시설 등 다른 프로그램들과 어떠한 방식으로 관계를 맺으며 도시조직에 삽입되고 증식될 수 있는지, 또 시간과 상황의 변화에 따라 어떻게 변형될 수 있는지 그 시스템에 대한 탐구에 초점을 맞추어 프로젝트를 진행했다. 이들이 제안하는 정글짐형태의 3차원 그리드 시스템은 마치 젠가 게임처럼 각각의 프로그램 모듈들이 필요에 따라 삽입되거나 제거될 수 있는 플랫폼으로서 기능한다. 3차원 그리드 구조를 따라 간단한 작물들이 재배될 수 있는 Green network가 형성되도록 하였으며, 좀 더 본격적인 도시농업을 위한 공간은 젠가시스템의 모듈로서 삽입되도록 제안되었다.

JANG Suho

Transmedia is the technique of telling stories or story experiences which can be understood as 'one' story through various media platforms. Transmedia storytelling is the text distribution custom and media experience pattern which has newly appeared in the age of digital convergence. Transmedia storytelling, a form of multiplatforming, makes a whole story by collecting individual stories. The Biodigital City Seoul Project induced construction companies to realize the final results of new buildings through transmedia storytelling. It gives you a reason to get closer to the future of urban agriculture by looking at the big picture of what is to come in the future.

The Bio Belt City of the A1 Team, suggested the direction of future urban agriculture through a story about a housekeeper's daily life. This was made to create an idea about the future of urban agriculture closer to you by establishing a connection to the housekeeper's life in a realistic setting with an ordinary character in the unrealistic urban agriculture form of the future. The Striped Farming City of the A2 Team analyzed the necessity of future urban agriculture and suggested a realistic direction for the future of urban agriculture by focusing on the present site and applying detailed simulations of future urban agriculture. The A3 Team tried to change their Jenga City into an urban agricultural space which included landscaped environments by overcoming simple agricultural functionalities and suggesting variable urban agricultural spaces of various types where future urban agriculture can be accomplished in building.

Through this, the Biodigital Seoul Project suggested the future direction of urban agriculture by presenting Seoul urban agriculture through transmedia storytelling in a familiar and creative way and inspired the realization of the future urban culture by producing videos that ordinary people and architectural experts can easily understand.

여러 개의 미디어 플랫폼을 통해 '하나'로 이해될 수 있는 이야기를 전달하고 이를 경험하는 것을 트랜스미디어라고 한다. 트랜스미디어 스토리텔링은 디지털 컨버전스 시대에 새롭게 등장한 텍스트 유통 관습이자 미디어 경험 양식이다. 멀티플랫포밍의 한 형태인 트랜스미디어 스토리텔링은 개별 이야기들이 모여 하나의 전체 이야기를 만들어내는 것으로 바이오디지털시티 서울 프로젝트에서는 트랜스미디어 스토리텔링을 통한 새로운 건축물의 최종결과물을 구현하도록 유도하였다. 미래 도시농업이라는 하나의 주제를 가지고 다양한 스토리텔링을 이용하여 앞으로 다가올 도시농업의 큰 그림을 그려봄으로써 미래 도시농업에 한발 다가서는 계기를 만들어 냈다.

한 주부의 일상을 통해 미래 도시농업의 방향성을 제시하는 A-1팀의 Bio Belt City는 자칫 먼 미래에 오는 비현실적인 도시농업의 형태를 평범한 주부라는 현실적인 캐릭터를 설정하여 주부의 일상 속에서 미래 도시농업이 친근하게 다가올 수 있도록 제작하였다. A-2팀의 Striped Farming City는 미래 도시농업의 필요성을 분석하고 현재 사이트의 모습과 미래 도시농업이 적용된 사이트의 디테일한 시뮬레이션을 중심으로 앞으로 다가올 도시농업의 현실적인 방향을 제시하였다. A-3팀의 Jenga City는 서사적인 스토리텔링이 아닌 한 건물에서의 미래 도시농업이 이루어질 수 있는 다양한 형태의 가변적인 도시농업 공간을 제시함으로써 단순히 농업의 기능성에서 벗어나 조경적인 환경을 포함한 도시농업공간으로 탈바꿈을 시도하기도 하였다.

이렇듯 바이오디지털시티 서울 프로젝트에서는 트랜스미디어 스토리텔링을 통해 다가올 미래의 서울 도시농업을 친근하고도 창의적인 방식으로 미래 도시농업의 방향성을 제시하고, 일반인부터 건축 전문가까지 쉽게 미래 도시농업을 이해할 수 있는 영상물을 제작하여, 앞으로 다가올 미래 도시농업의 실현 가능성을 고취시켰다.

BIO BELT CITY

KIM Bokyeong LEE Seokyoung LEE Youjin OH Byeonghoon

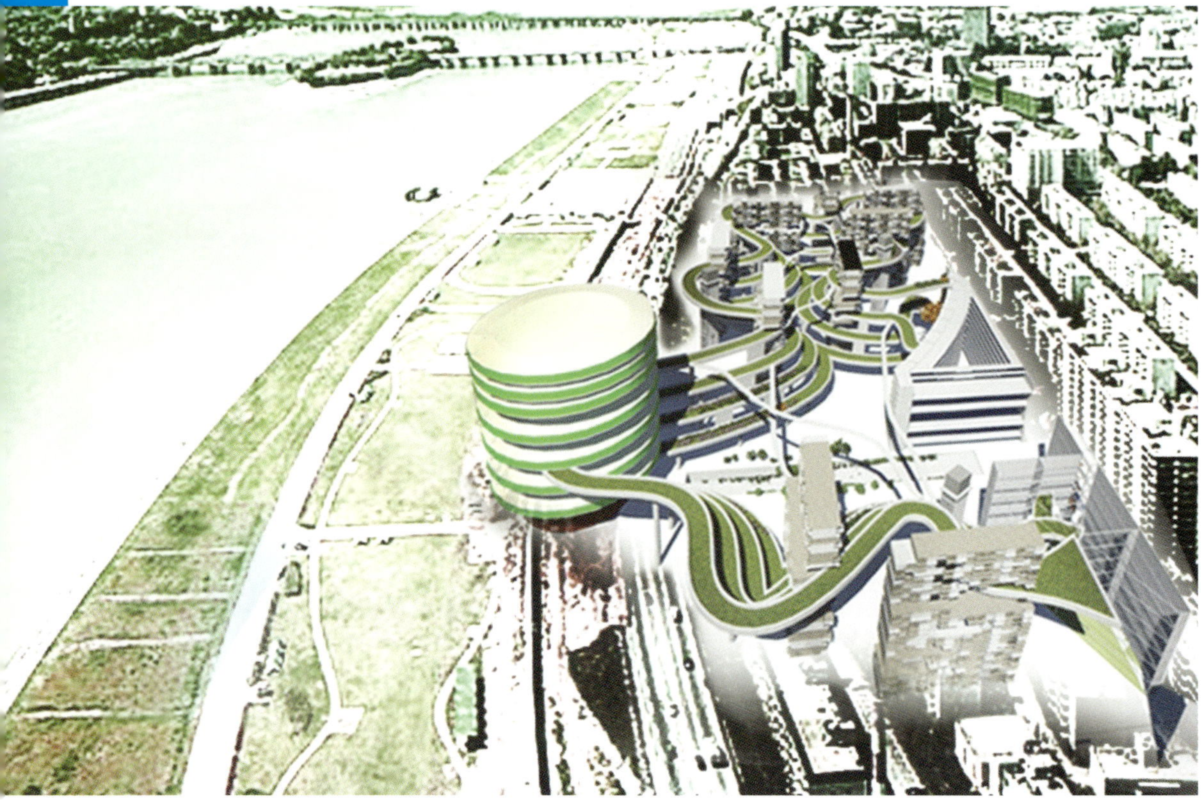

Bio Belt City

Ichon-dong, Seoul / Strategy / Perspective

How can 'cities' combine with 'agriculture' while these two seem so different by their very nature? The agriculture in urban life we suggest is not one of those one-time events such as weekend farms located outside of your daily routine, however, it is an alternative to regain the warmth of 'living' which has been forgotten in mechanized and individualized urban areas. This new agriculture form's goal is to let people in the city feel the presence of 'life' within their reach. Then how can this agriculture in an urban area penetrate into people's daily lives? Can we possibly add plant (agriculture) life cycles into daily urban routines to start? The 'Bio Belt' concept as a green blood vessel to connect whole residential complexes as one organic body started from there.

전혀 섞일 수 없을 것 같은 '도시'와 '농업'이 과연 어떻게 만날 수 있을까? 우리가 제안하는 도시생활에서의 농업은 일상을 벗어나 이루어지는 주말농장과 같은 이벤트성의 개념이 아닌, 기계화, 개인화 되어가는 도시생활에서 잊혀져 가는 '생명'활동의 따스함을 다시 느끼게 해주는 대안으로서의 농업으로, 도시농업을 통해 도시인들이 '생명'의 존재를 손 닿는 거리에서 일상적으로 느끼며 함께 살아갈 수 있도록 하는 데 목표를 둔다. 그렇다면 농업이 도시인의 일상생활에 자연스럽게 스며들 수 있는 방법은 무엇일까? 식물(농업)의 라이프사이클이 도시인의 라이프사이클에 들어와 하루 일과를 함께 한다면 가능하지 않을까 라는 상상을 시작으로, 건물 사이사이를 통과하며 단지 전체를 하나의 살아 숨쉬는 유기체로 만들어 줄 green blood vessel의 개념인, 'Bio Belt'의 컨셉이 생겨나게 되었다.

Living Complex with Bio Belt

xisting Site Condition / Mass tudy of Apartment Complex / Diagram

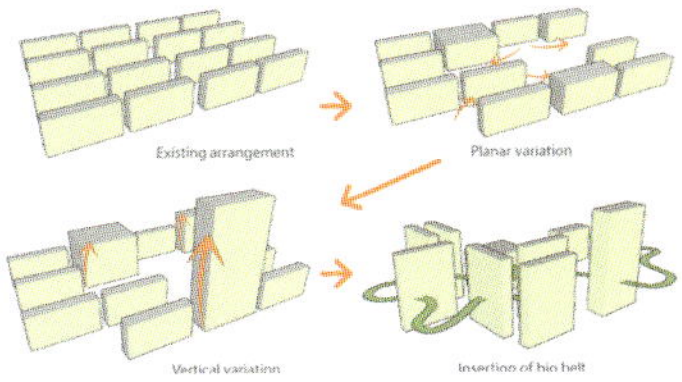

The Ichon site is located at the linear extension of the green network, and it is also a landmark that faces the Han River, but the current apartment complex is surrounded by walls and is also next to the Gangbyeon Expressway which makes the residents passive. Also, the separated structure of the apartment complex makes communication between residents difficult; not to mention the external green spaces between buildings are hard to access. So practically the only external space the residents can utilize is their balcony. On the other hand, the side of the complex that is opposite to the Han River is filled with stores and shops. Considering all of these aspects together, the new landmark town that is to be proposed as 'urban agriculture' is applied. The shopping district for both residents and visitors will be extended up to the external public spaces within the complex so that the public space will be naturally connected to the Han River park. The goal of this 'residential complex where agricultural and urban life cycles coexist' is added on the top of it to complete the new land -mark in Seoul as all of the houses, the neighborhood, stores, offices and public areas are connected via the Agricultural Bio Belt.

'The combining of city and agriculture' starts at the point where buildings (mass) and the Bio Belt meet. How much can the lives of people in urban areas be diversified by the conjunction? The focus of our design was to suggest this diversity. We brainstormed conceptual features of Bio and Digital together to imagine the various systems and spaces of urban agriculture. So the fundamental assumption of a Bio Belt is the presence of lines connecting across buildings of a complex to make the entire complex one town. The spots where the Bio Belt passes will apply various agricultural methods according to the program. Going further, combining the two physical concepts of Mass and Bio Belt, a totally new concept of space can be imagined and as a result, it will become an 'agriculture line' with possibilities to propose new urban agriculture.

이촌 사이트는 서울의 녹지축 선상에 위치해있고 한강을 면한 랜드마크로서의 장소성을 지닌 곳이다. 그러나 현재의 아파트단지는 한강쪽은 강변북로, 양 옆은 담장으로 둘러싸여 있어 한강이나 주변 지역에 대해 소극적인 자세를 취하고 있다. 또한 단지 내에 각각 독립적으로 분산되어있는 아파트 동의 배치구조는 동간 사람들끼리의 소통이 어려운 조건을 만들며, 동과 동 사이 지면에 위치한 외부녹지공간은 아파트 내 사람들이 쉽게 이용하기에 접근성이 좋지 않아 사실상 주민이 이용가능한 외부공간은 베란다가 전부인 상황이다. 한편, 한강 반대쪽의 면은 긴 상가건물이 위치해있어 지역의 생활가로의 역할을 하고 있다. 이러한 사이트의 여러 조건들을 바탕으로 '도시농업'을 접목시킨 서울의 새로운 랜드마크 타운을 계획하고자 한다. 단지 주민과 외부 주민들이 함께 이용할 수 있는 상가가 단지 내 외부 공공공간까지 이어지고 자연스럽게 한강공원까지 공공공간이 확대되는 성격을 지닌 복합주거단지를 계획하였다. 여기에 '농업과 도시인의 라이프사이클이 공존하는 주거단지' 라는 목표를 더해 도시인의 하루 일과가 이루어지는 집, 동네, 상가, 사무시설, 공공공간이 Agricultural Bio Belt로 함께 연결되어 서울의 새로운 랜드마크가 완성되었다.

'도시와 농업의 만남' 즉, 건물(mass)과 Bio Belt가 만나는 접점에서 농업이 도시인의 생활에 얼마나 깊게 파고들어 삶을 다채롭게 변화시킬 수 있는지, 그 다양성을 제안하는 것이 우리 설계의 포인트이다. Bio와 Digital의 개념적 특성을 브레인 스토밍한 후, 이를 바탕으로 다양한 도시농업의 시스템과 공간을 상상해보았다. 따라서 Bio Belt의 기본 전제는 단지 내 건물들을 뚫고 지나가 단지 전체를 하나의 복합 타운로 연결해주는 라인으로서, Bio Belt가 지나는 곳의 프로그램 성격에 따라 다양한 농업 방식을 적용하도록 한다. 뿐만 아니라 mass와 Bio Belt라는 물리적인 두 개념의 만남을 통해 새로운 개념의 공간을 상상할 수 있으며, 그 결과 새로운 도시농업의 생활을 제안해줄 수 있는 가능성을 지닌 'agricultural line' 이 될 것이다.

Bio Belt City Render

Various Agricultural Program / Perspective

Helping to imagine this new life style with a Bio Belt, we created a story about a house wife. On the key map, her moving line is presented and simple diagrams will be shown to introduce the bio or agriculture program at each space. And then a 3D rendered shot follows.

A day in the life of a house wife with urban agriculture is as follows:

1. Waking up - She wakes up looking at the plant on the wall and the ceiling extending from the veranda. While drinking water from the water purifier, a sensor activates to automatically water the plants as well. Breakfast is made with fresh ingredients from her own farm.
2. Morning - She goes out of the house to see her child off to school and passes through the Farm Bridge which is a corridor that connects each residential complex. Under the bridge, there's a duct with overnight rainwater stored in it which waters plants by itself. Along both sides of the bridge, there is a Community garden and elevator core that works as a bridge structure which takes people down to the ground. The elevator operates by solar energy.

Bio Belt가 단지 구석구석을 돌며 만들어내는 도시인의 새로운 life style을 상상하기 위해, 주부의 하루 일과를 따라다니는 스토리를 구상하였다. 간단히 도식화 한 key map을 통해 주부의 이동 동선을 먼저 보여준 후, 각각의 공간에서 실질적으로 이용되는 bio 혹은 농업 프로그램을 간단한 다이어그램으로 표현하고, 3d 렌더샷이 나오는 순서로 구성하였다. 도시농업과 함께 하는 주부의 하루일과는 간단히 다음과 같다.

1. 기상 - 아침에 일어나면, 베란다에서 자연스럽게 연장되는 벽/천장 작물을 보며 잠에서 깬다. 정수기를 통해 물을 마시면, 센서가 작동해 식물에게 함께 물을 준다. 직접 키운 식재료를 바로 따 신선한 요리를 할 수 있다.
2. 오전 - 아이를 배웅하기 위해 집을 나선 주부는 각 단지가 연결된 복도인 Farm Bridge를 지난다. 이 다리 밑으로는 밤새 모아둔 빗물이 이동하는 duct가 함께 지나며 자동으로 식물에 물을 공급한다. Bridge의 양쪽에는 단지내 사람들이 함께 키우는 Community Garden이 있고, 다리의 구조 역할을 하는 엘리베이터 코어를 통해 지상으로 내려간다. 엘리베이터는 태양광 에너지를 이용한다.

Bio Belt City Render

Various Agricultural Program / Perspective

3. Lunch - Arriving at the commercial building by passing through the elevator and stair-case farm, the owner of a restaurant cooks with fresh ingredients from a vertical farm. The vertical farm also maintains the privacy of the residents by reasonably covering the residential areas not to mention the beautiful scenery it provides.
4. Afternoon - She visits the Farm tower near the Han River with a child from school. As a landmark, it is located on the Han Riverside and the curtain wall building has professional programs, such as a vertical factory and R&D centers in it, which let children explore the agriculture while also providing a view of the Han River from the lounge.
5. Evening - The whole family goes out to the Han River park after dad gets home from work. The low accessibility to the park from the Gangbyeon Expressway is resolved with a wide underpath that expands the public area until it reaches the park. Both sides of the path are the underground space of the tower. A rice farm is also located within this space to supply soft street light at night not to mention the beautiful scenery it also provides.

3. 점심- 엘리베이터를 넘어 계단식 경작지를 지나 식당과 까페가 모여있는 상업 건물에 다달으면 식당의 주인은 수직농장의 식재료를 바로 따 신선한 요리를 해준다. 이 수직농장은 상업공간이 주거단지내에 있는 것을 고려해 외부인들로부터 주거공간의 프라이버시를 적당히 가려주는 역할을 하며 미관상의 이점도 지닌다.
4. 오후- 학교를 마치고 돌아온 아이와 함께 한강변에 위치한 경작타워를 방문한다. 랜드마크의 장소성에 맞게 한강변에 위치한 타워 형식의 건물로서 커튼월로 된 건물에 수직형 공장, 연구 등의 전문적인 프로그램을 지닌 곳으로, 아이들에게 농업에 관한 견학을 경험하게 해주며 한강뷰를 제공하는 라운지로서 또한 역할을 한다.
5. 저녁- 퇴근한 아빠와 함께 가족은 경작타워 밑에 위치한 터널을 통과해 한강공원으로 나간다. 이는 기존에 강변북로로 인해 한강공원으로의 진입성이 굉장히 낮았던 것을 보완해주는 것으로, 단지 전면이 한강공원에 접한 이점을 살려 단지 내 공공공간이 한강공원까지 확장되도록 하는 넓은 Underpath를 가진다. 이 길의 양옆은 타워의 지하공간으로, LED를 이용한 벼농장이 있어 지하에서도 경작을 할 수 있으며, 터널을 지나가는 사람에게 가로수와 같은 미관상의 역할뿐만 아니라 밤에는 은은한 가로등의 역할도 한다.

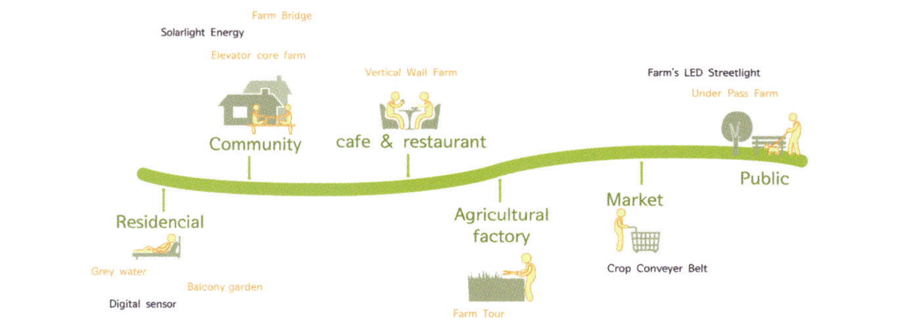

Intertwining Bio Belt

Insertion of Agriculture to Apartment / Bio Belt + 24hr City Life / Diagram

Bio, Digital and Farm

Strategry for New City's lifestyle with Agriculture / Diagram

Bio Belt City Map

Key Map / Diagram

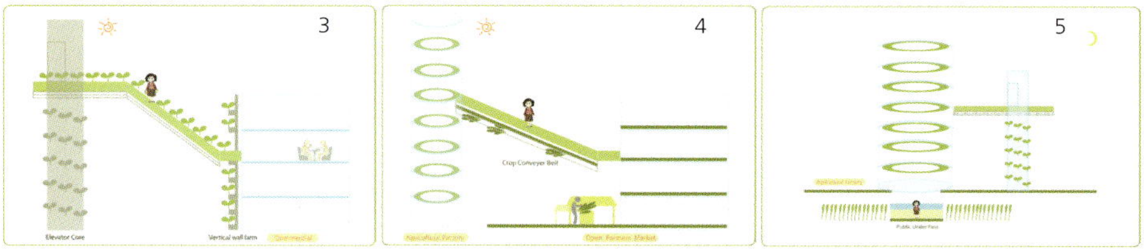

Bio City Life

Various Agricultural City Life / Keymap / Sectional Diagram

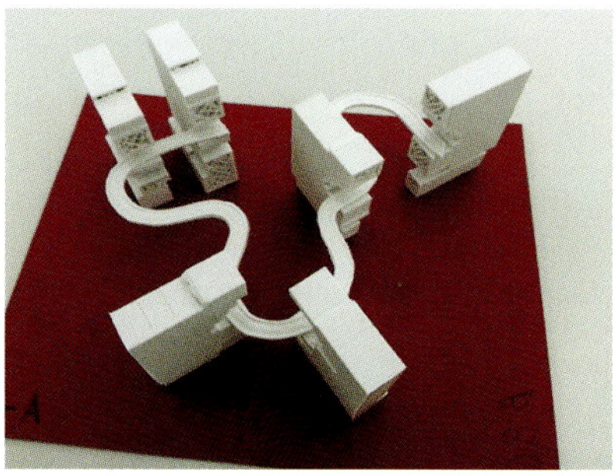

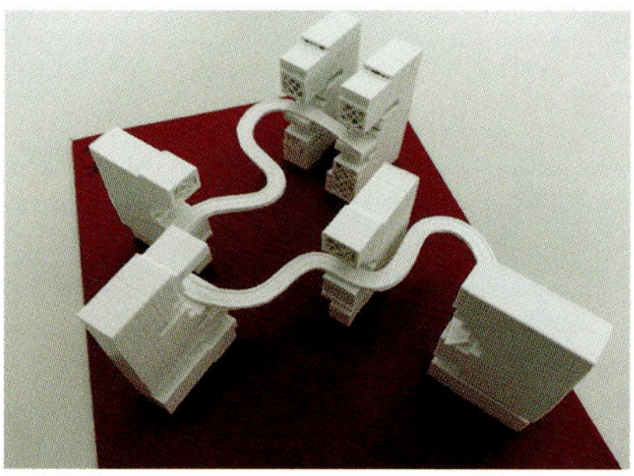

3D model

3d Printing /
20cm X 20cm X 7.3cm

My team designed the whole site, but due to time limits, we only printed the parts that show our concepts most effectively. That's why we chose residential parts and bio belt connections. The buildings were supposed to be filled up from the bottom with empty parts in the middle, so their various form were difficult to be expressed through 3D printing. the Bio belt was originally designed to be composed of flexible curves crossing through buildings but with 3D printing limitations, the outcome was presented as a flat curved belt.

The biggest issue we faced with 3D printing was time. We needed a model to present the concept within tight time limits. It was our intention to show how a 'green belt' crosses through the buildings and floats in the air. In order to present the 'belt' we imagined, each part should be separately printed. With the time restriction, however, we couldn't help but make a model that only showed our vision at the conceptual level. The issue with time continued throughout the project. While sharing a 3D printer with many other teams, the overall model should have been produced in a short period of time. But during the planning, good ideas kept coming out in the team and this resulted in delays deciding upon the mass shape. So the 3D printing was done with the deadline right around the corner. Also, the unexpected limitation of materials, colors, solidity and printing time was realized. There was a lot of trial and error but the model eventually described our intentions well enough.

Producing a model in architecture is a very important part of the process. If any changes are made to the design after the modeling, it needs to be done all over again which could cost a lot of time and increase expenses. That's why we needed to find solutions to resolve the problems that arose from frequent design changes. 3D printing is becoming more and more significant in that sense, it might be even considered essential to the field of architecture, to be able to design various forms without any restrictions. It was a priceless experience to embody the various creative designs of the students and also reduce the time and effort required.

우리 팀은 사이트 전체를 디자인하였는데, 전체를 모두 프린팅하기엔 많은 시간이 필요하기 때문에, 우리 팀의 컨셉을 가장 효과적으로 보여줄 수 있는 부분을 프린팅하였다. 그래서 주거부분과 바이오 벨트(bio belt)의 연결부분을 프린팅하였다. 건물의 경우, 중간이 뚫린 형태 등 다양한 형태로 3d 프린팅이 밑에서부터 채워가는 시스템으로 되어 있어 뚫린 형태를 표현하기 어려웠다. 바이오 벨트(bio belt)의 경우, 본래 유연하게 휘어져 건물의 사이사이를 통과하는 것으로 디자인되었지만, 3d프린팅의 기능면에서 한계가 있어 평평한 곡선형태의 belt로 나오게 되었다.

3d 프린팅을 하면서 가장 큰 문제였던 것은 시간이었다. 짧은 시간 내에 우리가 보여주고자 하는 개념을 쉽게 설명할 수 있는 모형이 필요했다. 우리 팀의 개념 특성상 'green belt'가 건물 사이를 통과하면서 공중에 떠 있는 형상을 보여주고자 하였다. 그러나 우리가 생각하는 'belt'를 정확하게 표현하기 위해서는 부재 하나하나를 프린팅하여 모형을 완성시켜야 했다. 그러나 시간적인 여유가 없었기 때문에, 우리는 개념적으로만 보여주기 위한 모형을 만들 수 밖에 없었다. 프로젝트를 진행 중 시간에 대한 문제는 계속되었다. 많은 팀이 3d 프린트를 이용하기 때문에, 빠른 시간 내에 전체적인 모형이 나와야 했다. 하지만 계획설계를 진행중에 팀원 간의 좋은 의견이 많이 나와 매스 형태를 결정하는 데에 많은 시간이 걸렸다. 결국 마감을 코앞에 두고 3d 프린팅을 하게 되었다. 또한 모든 것을 프린팅할 수 있을 것 같았던 3d 프린터는 우리가 생각하고 있었던 것보다 훨씬 많은 한계가 있었다. 프린터 출력시간, 재료종류, 재료색상, 재료강도가 문제점이 되었다. 프로젝트를 진행하면서 많은 시행착오가 많았지만 3d 프린터로 만든 모형이 우리가 말하고자 하는 내용을 잘 표현해 주었다.

건축에서 모형을 만드는 것은 굉장히 중요한 일이다. 그리고 건축 설계 과정에서 모델 제작이 끝난 후 설계 변경이 일어나면 다시 만들어야 한다. 그 때문에 시간과 비용의 소모가 커질 수밖에 없어, 빈번한 설계 변경으로 인해 발생되는 문제점들을 해소할 수 있는 방법이 필요하다. 그에 따라 3d 프린터의 필요성은 점차 중요해지고 있으며, 건축분야에서 많은 형태로 자유롭게 디자인하는데 필수적이라 할 수 있다. 건축 모델을 만들기 위해 학생들이 쏟아야 하는 노력과 시간을 줄이고, 학생들의 다양한 창의적인 디자인을 그대로 뽑아 낼 수 있는 하나의 소중한 경험을 하였다.

STRIPED FARMING CITY

Emilien GOHAUD LEE Seoyoung OH Hana

Striped Farming City

Seoul International Workshop / Ground Perspective

Do you want to eat fresh local food in the city? Live closer to nature? The striped farming city introduces a farming loop that will provide residents with quality food as well as a more diverse, fun, and eco-friendly living environment.

In the 1970s, the area south to the Han River began to be transformed into the most developed area in Seoul through mass production of apartment complexes, including Apgujeong Hyundai Apartment. While the area is still known for being the richest in Seoul, even after forty years, the actual living quality in the Hyundai Apartment Complex does not live up to its famed expectations despite its presence in a geographically exceptional city center location. Today, the site calls for renewal. Our group, rather than taking the traditional method of completely deconstructing the old buildings, took a new approach to enhance the sustainability by remediating part of the site's reusable housing units, a method which will transform the site's lethargic ambience into a diverse lively one.

도시에서 자연을 느끼며 신선한 로컬 푸드를 드시길 원하십니까? Striped Farming City가 거주자들에게 다양한 양질의 음식을 제공할 새로운 경작 시스템을 제안합니다. 자동화 시스템뿐만 아니라 생동감 넘치고 재미있는 친환경 라이프스타일을 즐겨보세요!

1970년대, 한강 남부는 서울에서 가장 발전된 지역으로 변모하며 압구정 현대 아파트를 포함한 대단지 아파트가 들어서게 된다. 현재까지 이 지역은 서울에서 가장 부유한 동네로 손꼽히며, 압구정 현대 아파트는 지리적으로 우수한 요건을 갖추었음에도 불구하고 거주 환경은 그 명성에 맞지 않는 실정이다. 따라서 이 지역은 재개발이 필수적이다. 이에 우리는 전면철거 후 재건축을 하는 관습적인 방식보다 단지의 일부분을 재사용하여 지속성과 다양성을 갖출 수 방법을 채택했다. 이 방법은 이 지역에게 보다 생명력 있고 다양한 활력을 불어 넣어줄 것이다.

Our urban attitude is key to this project; that not only individual apartment buildings, but also the space in between the buildings, such as the road access to both the river and mass transit, the parks and gardens alongside the residential sections, and regional commercial areas, were taken under consideration altogether. Part of our major change includes a new park that covers the highway, linking the city back to the river; a road scheme rearranged for better accessibility; a number of new commercial complex buildings which will be a crucial junction point where all the farming robots assemble together as part of the local farming-supply system. The robots, which will be explained shortly, follow along this extended linkage of vertical system atop the new commercial buildings and horizontally to the existing apartments, eventually creating a "One System" hybridized farming city.
On the southern façade of the existing apartments, linear stacked gardens are added to bring nature into everyday life. The linear farming runs all around the Hyundai apartment site in a striped pattern, horizontally from and vertically to the commercial complexes. This balcony-like gardening space serves both dietary and aesthetic purposes, either as a personal farming space or simply as a gardening space, depending on the preference of the person residing in the apartment. Multifunctional gardening robots slide along the stripes, the outer surface of the garden, and manage the crops on behalf of the residents. These stripes also work as solar panels that provide the power needed to run the robots.

우리의 도시활성화 방법은 개별적인 아파트 건물뿐만 아니라 건물들 사이의 공간, 한강으로의 접근성, 대중교통 이용의 편리성, 거주자의 편의, 상업시설까지 모두 고려되어 계획하였다. 주요 변화로는 한강변 고속도로를 공원으로 덮고 단지 내 도로를 재정비하여 사람들이 한강으로 다가가기 쉽게 하여 접근성을 좋게 하였다. 새로운 상업시설 단지는 지역 전체의 자동화 경작 로봇 시스템이 최종적으로 도착하게 되는 지점으로서 중요한 교차점과도 같은 공간이다. 로봇 시스템은 다음에 좀 더 자세하게 설명될 것이지만, 기존 아파트를 수평적으로, 새로운 상업 빌딩의 수직의 로봇 이동통로를 따라 움직이게 되어 통합적인 '원 시스템'의 농업 도시를 가능하게 한다.
기존 아파트 남측면의 확장된 테라스에 층층이 정원을 길게 설치하여 일상 속에서도 자연을 느낄 수 있게 하였다. 이 경작용 정원은 수평적으로는 단지와 단지를 연결하며, 수직적으로는 상업시설로 연결되어 지역 전반을 리본처럼 감싸는 형태를 가진다. 이 공간은 생산적, 심미적 두 가지 목적을 모두 충족시키며 거주자의 취향대로 개인 정원 공간 또는 경작 공간으로 사용될 수 있다. 다목적의 경작로봇은 테라스 외부 레일 시스템을 따라 이동하며 바쁜 현대인들을 대신하여 식물 재배를 돕고 수확을 한다. 또한, 레일 표면을 따라 설치된 태양 전지판은 로봇 가동에 필요한 전력을 제공하도록 설계되었다.

Striped Farming System

Seoul International Workshop / Automatic Striped farming system / Diagram

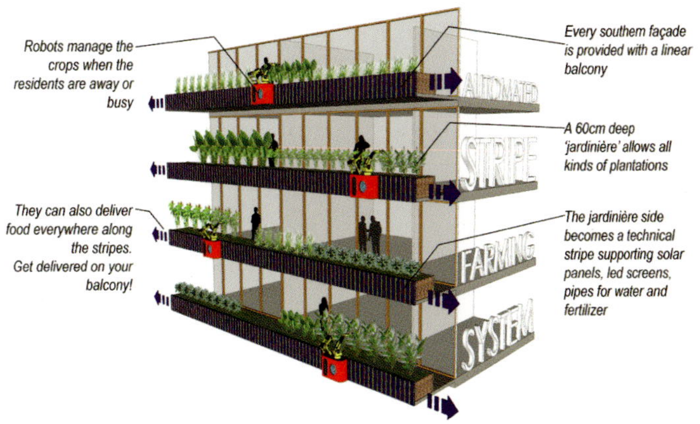

Striped Farming City

Hybridized City / Global View

Every old and new, small and big building is hybridized by the striped farming system. This connection also serves as a metaphor of how the old and the new integrate harmoniously in a green space to which human beings has aspired to for so long. Besides having implicit meanings, the robot lane connection also has practical functions. Running down the vertical tower, the stripes meet at a street-leveled lower building that intersects with the apartments. Through this extended linkage from the residential blocks to the commercial buildings, the food produced is also served to the public in local restaurants and stores. Live in the dense city, with an easily manageable vegetable garden, and eat local veggies!

모든 옛 건물과 신 건물, 크고 작은 건물들이 스트라이프 패턴의 경작 로봇 시스템을 통해 통합된다. 선들은 수직농장타워를 지나 도로변 저층건물에서 만나고, 아파트 건물과도 교차한다. 이 연결은 구 건물들과 신 건물들이 어떻게 조화롭게 통합되는지를 보여주며, 현대인의 오랜 염원이였던 자연과 함께 하는 삶을 가능케 한다. 뿐만 아니라 신(新) 구(舊)의 '연결' 이라는 상징적인 의미를 넘어서서, 이 로봇 시스템은 실제로도 옛 건물과 새 건물의 연결고리 역할을 한다. 층층이 겹쳐진 스트라이프 패턴의 경작 로봇시스템은 구(舊) 아파트에서부터 연결되어 신(新) 수직 타워를 따라 내려가, 아파트 가장자리에 직각으로 이어진 저층 신(新)건물로 확장 연결된다. 이와 같이 주거시설과 상업시설을 거시적으로 연결함으로써, 아파트에서 길러진 작물이 지역 레스토랑과 가게에도 제공 가능케 한다.

Rearranging

Road Works & Builidng Removal & Afforestion / Map / Diagram

Road Works & Builidng Removal & Afforestion / System-in-Zoom / Diagram

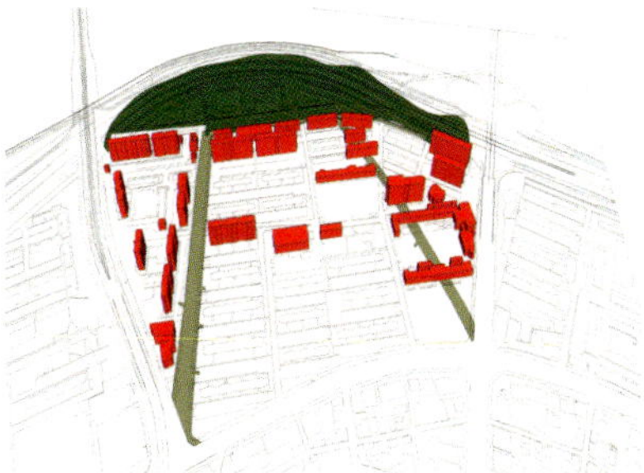

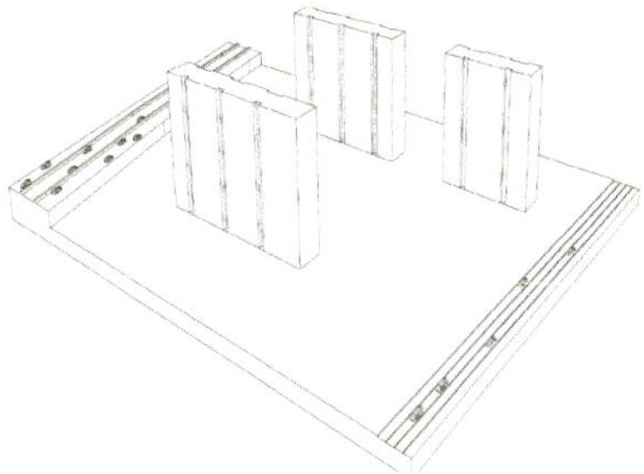

Adding

Added New Buildings / Map / Diagram

Added New Buildings / System-in-Zoom / Diagram

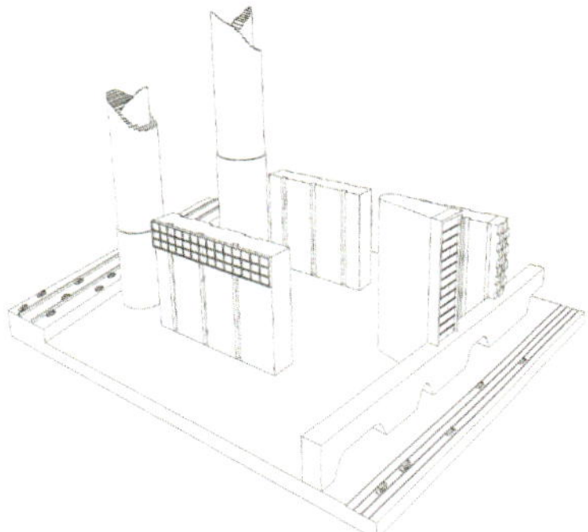

Linking

Striped farming system / Robot rail connection / Map / Diagram / Ontents

Striped farming system / Robot rail connection / System-in-Zoom / Diagram

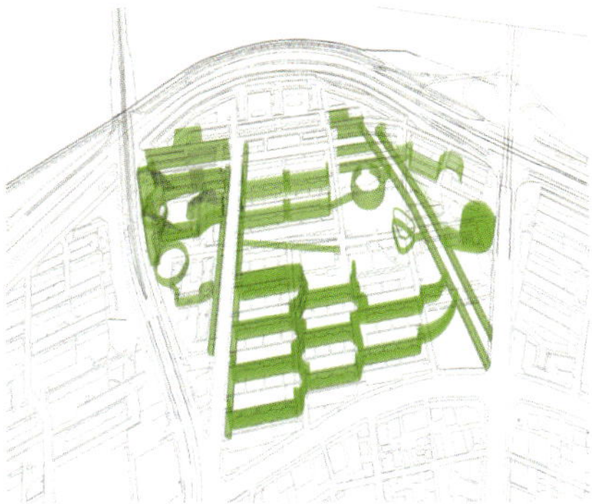

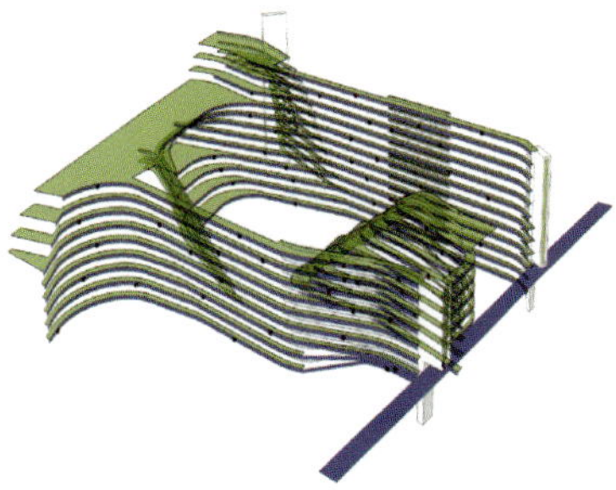

Hybridized City

Hybridized City with Striped Farming System / Map / Diagram

Hybridized City with Striped Farming System / System-in-Zoom / Diagram

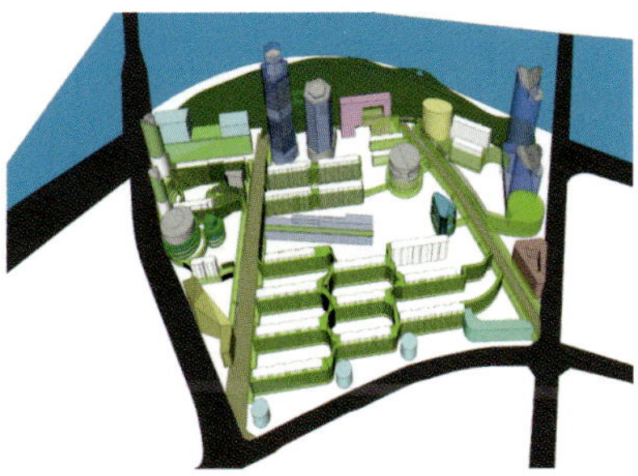

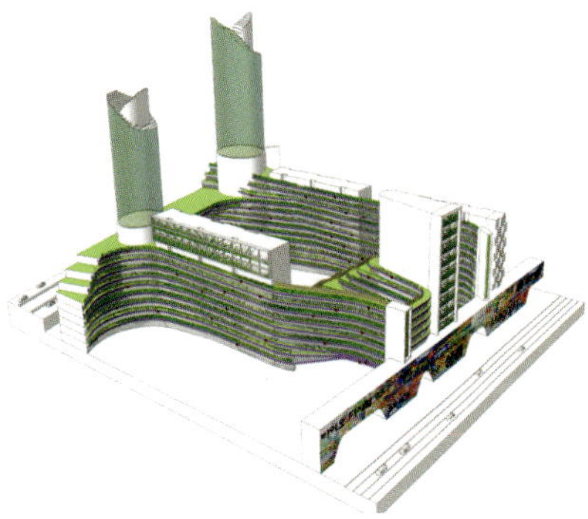

Striped Farming City_ Community Area

Community Area / Perspective

New Life Style With Striped Farming City

Change in Life Style suggested by Striped Farming City / Diagram

Apgujeong Hyundai Apartment is a dull housing complex associated with an unhealthy rat-race lifestyle. The stripe farming system triggers a radical change toward a more natural lifestyle. The existing site is progressively transformed and integrates the striped farming system, resulting in a complex, fun and productive urban environment.

Our team was composed of members specialized in architecture and urban planning, but none was familiar with media. Due to the absence of a media art expert, we faced our greatest challenge to figure out the most effective way to represent our idea visually in 5 minutes. After going through several tests and trials, we decided to take the simplest but the most effective method, to show a virtual walk-through of our 3D modeling, along with some descriptive captions as the scenes flow by. Though we did not have any extravagant visual effects or state-of-the-art camera techniques, we managed to show the majority of our important concepts by turning frame to frame that which was taken from certain vantage points.

압구정 현대아파트는 칙칙하고 획일적인 주거단지로, 거주자들에게 다람쥐 쳇바퀴 돌리듯 건강하지 못한 생활방식을 안겨준다. 그러나, 줄무늬 경작시스템을 설치함으로써, 거주자들은 더욱 건강한 생활방식을 얻고, 현대아파트 지역은 점진적으로 변화하여 더욱 다양하고, 즐거우며, 생산적인 도시환경을 제공한다.

영상 제작 전문가의 부재로 팀의 아이디어를 영상으로 효과적으로 보여줄 수 있는 방법이 가장 큰 고민이자 도전이었다. 여러 번의 시도와 테스트를 거쳐 간단하면서도 가장 효과적인 3D 모델링을 통하여 걷는 시각적 효과방법을 사용하기로 결정하였다. 영상의 흐름과 함께 간략한 설명도 추가하였다. 우리는 화려한 영상 효과나 최첨단 카메라 기술 없이도 팀의 핵심적 아이디어를 가장 효과적으로 보여줄 수 있는 방법으로 중요한 개념을 전달할 수 있었다.

Working with 3D printing posed a set of challenges. It was a very interesting process to go through, and we were pleased by the quality of the final work yielded by the 3D printer. As we continued working with 3D printing, we learned about the limits of time and geometry that the 3D printer has, and even though we went through a lot of trail and error, we somehow reached the best state of what we could achieve.
What we were trying to show via compact 3D modeling was an integrated farming city of the future. In order to do so, we had to show all of the elements that we had, such as old and new buildings, commercial complexes, and smart robot systems where production, transportation, and consumption take place all at once.

In order to show existing buildings, new high rises, and commercial complexes combined within a robot system, all three types of building had to be printed. However, an error occurred due to the steep slope of the model, and even worse, we ran out of time and were unable to print everything. As a consequence, we printed individual elements separate from each other, and added them later on. Printing each element in a different color also let the viewer comprehend it better.

3D 프린터를 이용한 모형 제작과정은 호기심과 도전의 연속이었다. 매우 흥미로운 작업이였고 프린팅 과정, 결과물의 높은 완성도에 만족할 만한 성과를 이루어 뿌듯한 마음이 들었다. 동시에 3D 프린터의 시간적, 물리적 한계 역시 알게 되고 이 과정에서의 오류나 실수를 최소화하기 위해 많은 수정과 시도의 반복이 있었기에 완성도 있는 결과물을 갖게 되었다.
우리팀이 3D 모델로 보여드리고자 한 것은, 통합된 농업 미래 도시를 컴팩트하게 나타내고자 함이였다. 그리하여 새로운 복합 고층 빌딩과 기존 아파트건물, 상업시설이 모두 스마트 로봇 시스템으로 연결되어 생산, 운반, 소비가 동시에 이루어지는 원 시스템을 제시하였다.

기존건물, 새로운 고층빌딩과 상업시설이 로봇 시스템 레일로 연결되어 통합된 모습을 보여주기 위해 3가지 건물이 모두 프린트되어야 했다. 하지만 3가지 건물이 연결되어 있는 모형을 3D 프린터로 표현되기에는 경사가 심해 프린팅 오류가 발생했고, 뿐만 아니라 전체를 한번에 프린트하기에는 시간이 너무 오래 걸리는 문제가 발생하였다. 그리하여 기존 건물, 새로운 건물, 로봇 레일 시스템 세가지 요소를 부분적으로 프린팅하기로 결정하였다. 이 방법은 여러 대의 3D 프린터를 이용해 동시에 다른 요소를 프린트함으로써 시간 단축에 용이했고, 오류가 발생했을 때 수정과정이 간편하였다. 또한 이 과정에서 3가지 요소를 각기 다른 색으로 표현하여 사람들의 이해를 높이는 모형 제작을 할 수 있었다. 따라서 기존건물은 흰색, 새로운 빌딩과 상업시설은 파란색, 통합 로봇 레일 시스템은 녹색으로 표현되어 팀의 '원 시스템 통합 농업도시' 아이디어를 가장 효과적으로 전달할 수 있었다.

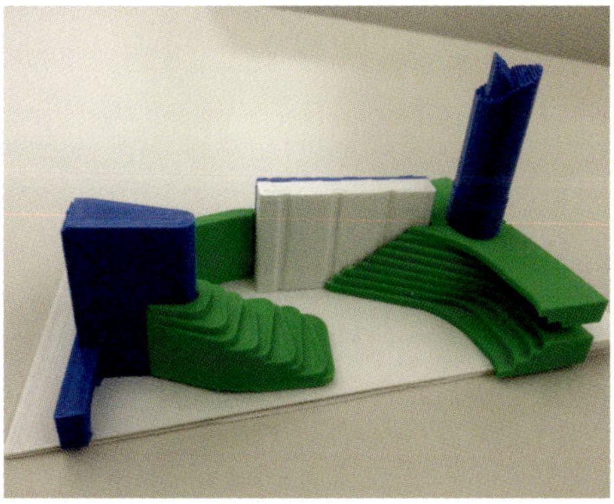

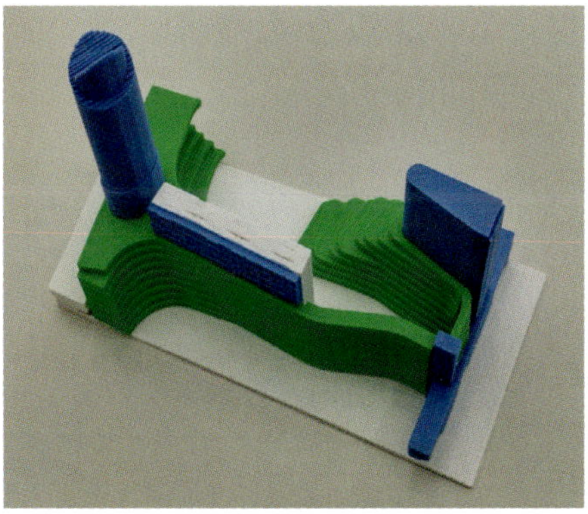

3D Model_Front

3D Printing / Front View / 15cm x 10cm x 9cm

3D Model_Back

3D Pringting / Reverse side View / 15cm x 10cm x 9cm

JENGA CITY

HONG Jiyeong KIM Hyunjin SHIN Hanna Vincent MARSAT

Jenga City

Jenga City Sky View / 14.08.2014

This project aims to build a necessary environment in order to induce the migration of people to an aging urban area and its cultivating activities. The site consists of old apartments named Hanyang Mansion, in Ichon, Seoul, which are surrounded by high-rise buildings that have been redeveloped. The existing low-rise apartments fail to achieve their potential in this area to accommodate more people like the surrounding new buildings do. There is also a problem with the height contrast with all the surrounding buildings. Therefore, it is necessary to change the skyline to consider the surrounding high-rise buildings as well as to enable agricultural activities in the city.

Typically, the renewal process in South Korea requires evacuation of the inhabitants, but this causes a problem because there is nowhere for these people to go. We faced this problem and planned to enable people to build a new building themselves rather than following the conventional renewal process with the destruction of all existing structures.

빠른 기술의 발전과 편리한 생활의 증진으로 현재 우리가 살아가고 있는 도시와 우리의 몸은 병들고 지쳐가고 있다. 때문에 우리는 이동 중에 배출되는 탄소 배출을 줄이고, 화학 처리없는 신선한 야채를 먹고, 우리가 속한 자연의 원리를 잊지 않기 위해 노력할 이유는 충분하다. 하지만 이러한 기술의 진보는 디지털화된 사회를 친환경적으로 변화시킬 수 있도록 하였다. 예를 들어, 과거의 논, 밭에서만 농사를 지을 수 있다면, 이러한 신기술은 우리가 실외뿐만 아니라 실내에서도 경작이 가능하게 하였으며, 땅뿐만 아니라 하늘에서도 경작을 가능하게 하였다. 이러한 기술로 인해 현재 우리 사회에서는 다양한 형태로 생겨난 도시농업의 사례를 찾아볼 수 있다.

이 프로젝트는 이러한 다양한 사례들을 참고하고 발전시켜 이촌의 낙후된 땅에 거주하는 사람들로 하여금, 농경활동이 가능하도록 유도한다. 특히, 이촌에 위치한 한강맨션은 주변의 고층아파트로 둘러 싸여 있는 오래된 아파트 중 하나이다. 이러한 지리적 특성으로 인해 한강맨션은 주변과 대조되어 매우 낮은 건물의 형태를 가지고 있다. 그러므로 한강맨션은 주변의 스카이 라인을 고려해 변화되어야 할 필요가 있다.

Jenga City System

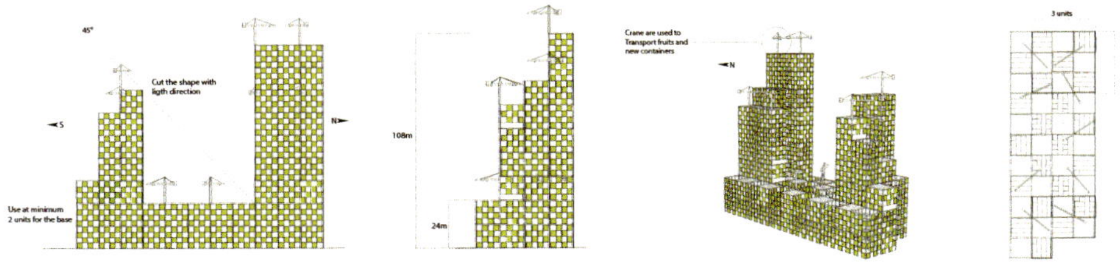

Based on this context, the project title is determined as 'Jenga City'. This came from the Jenga game that we all know. While there are wooden blocks stacked to create a high tower in the Jenga game, similarly there are recycled shipping containers vertically stacked in this project. Shipping containers usually last for about 15 years and then they are discarded, but the large volume and heavy weight cause problems, and they are treated as industrial waste. So another purpose of this project is to encourage people to reuse these shipping containers to protect the natural environment and at the same time to build accommodations with a low-cost budget. Additionally, it is possible to grow plants inside the container by changing the interior finishing materials to a compact assembly in which plants root. This enables people to cultivate inside the container as well as outside of it.

기본적으로 한국에서의 재개발시스템은 기존의 주민들을 모두 다른 곳으로 보낸 후, 재개발 공사가 진행된다. 하지만 이러한 시스템은 기존에 거주하던 주민들이 이주할 곳이 없다는 문제점을 만들기도 한다. 우리는 그러한 문제점을 직시하여, 기존의 건축물을 모두 붕괴시킨 후, 다시 새로운 건축물을 올리는 것이 아니라, 사람들이 필요에 따라 스스로 건축을 하여, 자체적으로 새로운 건축물을 확장시켜 나갈 수 있도록 계획하였다.
이러한 형태를 기반으로 프로젝트 개념은 'Jenga City'로 정하였다. Jenga City는 우리가 알고 있는 Jenga 게임의 형태를 말하는 것이다. Jenga 게임에서는 나무를 쌓아 높은 탑을 만들어 나가지만, 이 사이트에 새워질 새로운 건축물은 기존의 콘크리트 건축물이 아닌 폐컨테이너를 재활용하여 Jenga의 형태로 배치하여 사용한다. 폐컨테이너는 그 역할을 하는 수명이 15년 정도로 그 수명이 다하면 폐기되지만, 그 부피가 매우 크고, 무게가 매우 무거워 산업폐기물로 처리하기에도 매우 곤란하다. 이러한 폐컨테이너를 이용하여 자연을 보호함과 동시에 저렴한 비용으로 건축이 가능하도록 하는 것이 이 프로젝트의 또 다른 목적이다. 이러한 폐컨테이너의 내부는 기존의 재료를 식물이 뿌리를 내릴 수 있는 재료로 변경하여 컨테이너 내부 어디에서든 식물이 자라는 것을 가능하게 하도록 하여 컨테이너 내부 즉, 실내에서도 다양한 경작활동을 할 수 있도록 한다.

Concept &
Special System

Concept is 1+1=3 /
This city have smart appri-
cation for farming and live.

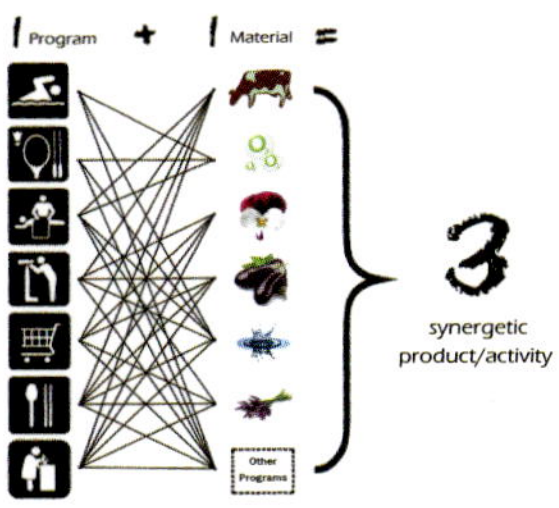

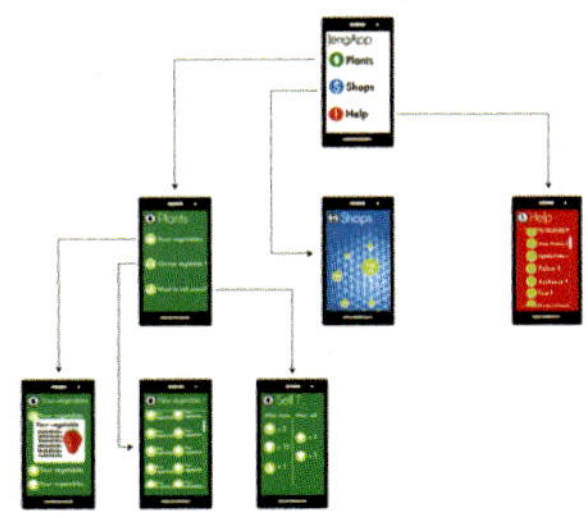

Image Sketch

This image is sketch for making video synopsis and concept.

A unit space like a container can evolve into various types of spaces that can then accommodate various programs. Also, the agricultural products produced by people can be not only consumed directly but also constitute a basis for further economic activities. Therefore we suggest establishing a strategy that creates a greater synergy (1+1=3 effect) by using a smart phone application.

A programed application will be able to provide people with the optimized information that they need to find locations to live, the type of economic activity (needed, available or on site) and the species of plants to cultivate. 1+1=3 means that we can produce a new program (3) by combining one present urban public program (1) and agricultural products (1) beyond theis simple combined effect. For example, if you grow herbs as a crop while operating a beauty shop, the beauty shop can become an eco-friendly laboratory which experiments with the use of herbs in an urban community. Such potential of synergy will encourage independent economic activities of the inhabitants and increase the identity of the community as a whole.

On the other hand, there is a second system besides the Jenga system which looks like a 'jungle gym' around the Jenga towers. It has been designed to be easily proliferate so that the system can cover the site with its structure. This 'jungle gym' consists of cubes which have 12 edges, each being 12 meters long, and each edge contains complete mechanical and electronic systems. The flowing water inside the pipe enables people to farm using hydroponics. Also, the 'jungle gym' structure can physically help the containers to be stacked higher by providing better stability.

Jenga 형태의 타워를 형성함과 동시에 타워 주변에는 정글짐 형태의 시스템이 계획되었다. 이 정글짐 시스템은 12m 길이의 구조설비용 각관(construction equipment squares pipe) 12개가 연결되어 정육면체를 형성하며, 그 범위를 넓혀 나갈 수 있도록 설계되어 있다. 이 파이프 내부는 전기설비, 수도설비, 가스설비 등 다양한 설비시스템이 들어감과 동시에 파이프 내부에 물이 흘러 파이프 자체가 수경재배를 할 수 있는 농업시스템으로도 사용될 수 있도록 하였다. 또한 이 각관은 컨테이너 박스의 구조적인 역할을 하도록 하여 컨테이너 박스가 더 높은 층을 형성할 수 있도록 구조적으로 도와주는 역할을 한다.

이 사이트의 모든 프로그램들은 사용자의 선택에 의해 결정되어지며, 사용자의 선택으로 인해 주변의 다른 환경적 요소 및 다양한 프로그램들이 결합되어 더 큰 시너지효과(1+1=3 System)를 만들어 내도록 한다. 예를들어 사용자들은 자신이 원하는 환경, 재배할 식물, 프로그램 등을 스마트시스템에 입력하면, 그 시스템에서는 주변의 기존 사용자들의 정보와 사이트의 특성, 환경 등을 분석하여 사용자가 원하는 조건을 갖춘 장소를 선별해 주고, 사용자는 그 선별장소를 선택하여 입주하도록 한다. 또한 1+1=3 System으로는 예를들어 입주자가 농작물로 허브를 키우는 것을 선택하고, 프로그램으로 뷰티샵을 선택하였다면, 이 두가지의 요소가 결합하여 친환경 허브뷰티샵이 생성될 수 있다. 이러한 시너지 효과로 인해 거주자들로 하여금 현 사이트의 특징을 더욱 높이도록 한다.

Jenga City Program

Jenga City have something program. For example, 'Shop, Housing, Farm, Etc.'

Consequently, Jenga City has been designed to become a high-rise building up to 40 floors considering the surrounding skylines, the higher part are separated into several parts so that they cannot disrupt each other in terms of access to sunlight. The Jenga City Project suggests, therefore, an alternative formation of living in an urban setting which is vivid, fun and high-density. There is a vegetable wall system riveted on the Jungle gym structure, which is installed in order not to shadow the crops. The height of the towers has been designed to maximize sunlight received from the south. Hopefully, Jenga City, located in the middle of Seoul, will become a symbolic role model of urban farming by developing itself as if it is a dense forest in the city. This will suggest the conviction that we can also cultivate in an urban environment.

마지막으로 이 사이트는 주변의 스카이라인을 고려해 최대높이 40층까지 가능하도록 설계하였으며, 고층부는 일조권을 고려해 몇 동의 타워로 분리하였다. 결과적으로 Jenga City Project는 기존 우리 사회의 획일화된 아파트 형식의 단조로운 주거형태가 아닌 더욱 생기있고 재미있는 고밀도의 주거형태를 제안한다. 정글짐 구조의 벽면에 설치되는 벽면 경작 생활시스템(farm living wall system)은 태양으로 인해 그림자가 생기더라도 정글짐 내부의 건물에서 일어나는 경작활동에 피해를 주지 않도록 설계되었으며, 건물의 높이는 남쪽보다 북쪽이 더 높은 층을 형성하도록 하여, 남쪽의 채광조건을 최대한 이용할 수 있도록 하였다.
이촌은 북쪽으로는 남산, 남쪽으로는 한강이 위치하고 있는 서울의 지리적 중심지이다. 이러한 장소에 도심 속의 숲처럼 형성될 Jenga City는 도시농업의 상징적 본보기가 될 것이며, 도심에서도 경작을 할 수 있다는 확신을 가져다 줄 것이다.

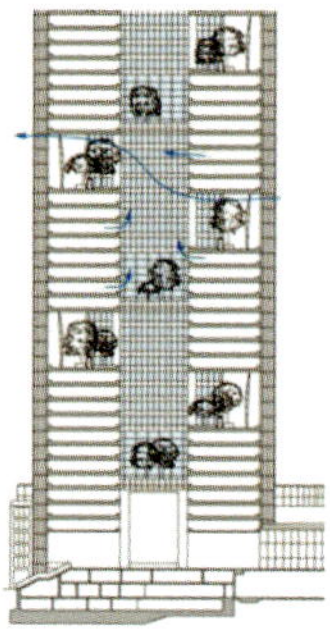

Commerzbank
Foster and Partners

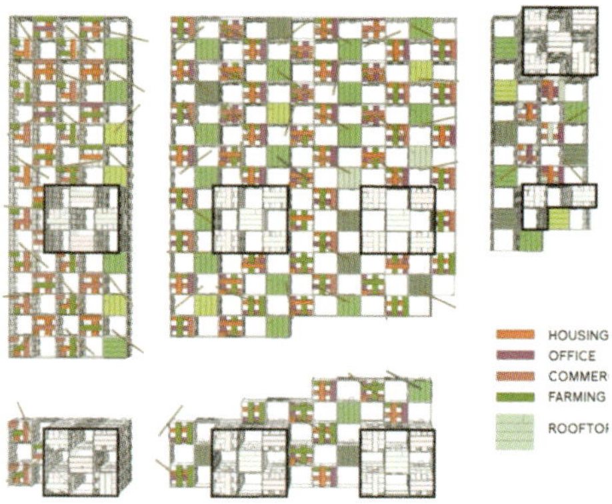

Rule For Self-Installation

Do not block the neighbors sunlight. Because this city have smart application.

The 'Jenga City' project was made by a 3D program from the beginning to the end.
The first scene of this movie starts with jenga virtually made by 'MAYA'. By disassembling a piece of jenga, the camera view zooms out to look down upon the whole city of Ichon, then it zooms in to our site, 'Han-gang mansion'. The following cloud of words are the keywords of our investigation about the site and the theme of this project. The concept of the movie is to explain how Jenga city replicates itself and has been made by using 'CINEMA 4D' with its short descriptions.
Getting into the main story, the movie shows the appearance of Jenga City with a birds-eye view. The virtual camera is continuously close to the frame of Jenga City, which functions as well as hydroponics. This scene has been animated very colorfully to show how the plants are growing under the sun and with water.
The interior composition is shown by a fastly moving virtual camera. The following conceptual diagrams about the masterplan and agricultural-economic programs was visualized by using 'After Effect'. At the end of the movie, the first scene is repeated but the piece is now reassembling the jenga tower.

영상작업과정 및 스토리 프로젝트 'Jenga City'는 처음부터 끝까지(3분 49초) 3D 프로그램으로 제작한 작품이다. 영상의 첫 장면은 3D 프로그램인 'MAYA'로 제작된 젠가로 시작한다. 젠가의 스틱하나가 빠져나오면서 용산 도시의 전체 이미지가 나오고 SITE로 지정된 한강맨숀을 극대화시키며 스토리가 전개된다. 이어서, 그 곳의 현황을 조사한 내용과 현재 디지털 바이오 시티의 핵심이되는 단어들이 연출된다.
본격적으로 하늘에서 땅을 내려다 보는 각도로 Jenga City의 외부의 모습을 보여주며 시작된다. CINEMA 4D(프로그램)을 이용하여 Jenga City가 형성되는 과정을 간략한 설명과 함께 보여준다. 건축 외형 이후에 내부의 모습이 서서히 클로즈업되면서 가상의 카메라가 돌아다니면서 찍는다. Jenga를 넣은 틀인 파이프의 내부 구성은 채소, 야채 등을 키우는 공간이 있다. 이 부분은 특히 Jenga City의 핵심으로 태양과 물, 그리고 자라나는 식물의 색감을 선명하게 연출하였다.
실내 구성은 가상의 카메라를 이용하여 이동하는 모든 경로들을 빠르게 화면전환하여 보여준다. 거주하면서 사람들이 할 수 있는 바이오 프로그램들과 마스터플렌을 영상 효과 After Effect(프로그램)을 통해 텍스트와 이미지함께 표현된다. 건축의 내부 단면을 잘라 전체 구조를 보여주며 영상처음에 나온 3D로 제작된 Jenga가 나오며 빠져나온 스틱이 다시 원상태로 돌아가면서 영상은 막을 내린다.

The 3D model of our design was digitally fabricated by using a 3D printer. In order to use the tool correctly and with ease, an expert gave a very detailed and careful explanation. Computational modeling in Rhinoceros was prepared to be 3D printed, and we decided to use only one color - white - for an objective evaluation of the printed model. The most important point of our making a 3D model was not to show a physical appearance but instead, to have people experience the Jenga Game with our 3D model so that they could understand our basic concept about how to develop Jenga City.

The resultant model of Jenga City consists of two parts. The pieces on one part are in the shape of a box which stands for the container, and the other part is the pipe structure that conveys the necessary environmental facilities. The boxes are 100 x 500 x 100 (mm) and engraved with their program titles such as HOUSING, SHOP and so on. The pipes are separately modeled and printed in two different types - joints and rods. Consequently, people could understand the basic concept of Jenga City by playing with the 3D printed model - the rod shaped pipes with joint pipes could be assembled in a Jungle Gym shape first, and then boxes fill the structure.

3D 모델의 기본적인 제작방식에 대하여 관계자분의 충분한 설명이 있었고, 팀 프로젝트의 형태가 비정형이 아닌 정형의 형태를 자지고 있었기 때문에, 3D 모델의 제작은 순조롭게 진행되었다. 기본적으로 모델의 인쇄를 위해 컴퓨터 프로그램은 라이노를 이용하여 진행되었으며, 3D 모델의 객관적인 평가를 위해 색을 입히지 않은 흰색만을 사용하여 모형을 제작하기로 결정하였다.

3D 모델 제작에 있어서 가장 중요한 포인트는 이번 프로젝트의 실질적 형태를 만드는 것이 아닌, 젠가라는 기본 개념과 젠가시티가 형성되는 방법을 체험할 수 있게 하는 것이다. 젠가시티의 기본형태는 도미노 블록의 형태 즉, 컨테이너박스와 컨테이너박스에 다양한 설비시스템을 공급해 주는 정글짐형태의 파이프였다. 이러한 형태를 기반으로 3D 모델 제작 작업은, 우선적으로 100X500X100(mm) 크기의 박스를 라이노 프로그램을 이용하여 디자인하고, 컨테이너 박스의 Program Name(Housing, shop)을 박스에 새겨 표시하였다. 또한 파이프와 파이프 연결고리를 각각 만들었다. 결론적으로 각각의 파이프를 연결고리에 결합시켜 정글짐 형태를 완성시킨 후, 정글짐 사이사이에 원하는 프로그램명이 적힌 컨테이너박스블록을 넣으면 젠가시티의 기본적인 개념을 이해할 수 있다.

3D Model

Making a Model using cinema 4D program.

T - B

MOON Junsik

This international workshop was different from other common architectural design workshops. It provided sufficient opportunities for participants to explore how urban farming can be applied to the metro city of Seoul, Korea, using the newest digital technologies. The participants of the workshop also had various backgrounds such as architectural design, engineering, urban design, and multi-media arts. Even though the workshop was unique and special, many things should be considered in many aspects as a tutor in order to make participants complete this project on time. Thus, I tried to do my best to attract the convergence of various creative and innovative ideas and thoughts from participants with a consistent view of the big picture of the final result. At the same time, I tried to foster efficiency since this workshop had very short time frame, less than seven days. For instance, the schedule was made based on the worst case scenario. In doing so, our teams were able to complete the project on time with great quality. I strongly believe that the workshop gave valuable experience to the participants and even to the mentors in terms of providing opportunities to look at problems from various points of view, to work with people who work in different fields, and to learn new media methodologies.

이번 프로젝트는 단순한 건축공간 디자인 워크샵이 아니라 최신의 디지털 기술들을 기반으로 도시 농업을 서울이라는 도심 공간에 융합해보는 실험의 장이었으며, 동시에 건축디자인 및 공학, 도시분야, 멀티 미디어 등 다양한 전공분야를 가진 학생들의 창의적이고 생산적인 아이디어들의 융합의 기회를 제공하였다. 하지만 짧은 기간 동안 다양한 생각과 백그라운드를 가진 학생들끼리 의견을 하나로 모으면서 효율적인 작업을 하는 데에는 어려움이 따를 것이라 예상하였다. 따라서 워크샵 시작 전에 각 조별로 구성원들의 관심사 및 장점을 파악하는 데 주력하였으며, 각 장점들을 모아 시너지 효과를 가져올 수 있는 문제해결을 위한 융합에 초점을 두고 지도하였다. 이를 위해 주어진 대지에서 해결해야 할 문제점들을 디자인, 공학, 미디어, 도시, 금융 등의 다양한 분야로 나누어 제시하여 각 조원들 개개인의 특기 및 장점을 발휘할 수 있는 여건을 제공하는 데 힘을 썼다. 가장 힘들었던 의견 조율부분에 있어서는 우리가 해결해야 할 가장 큰 목표를 항시 상기하도록 유도하였으며, 세부적인 분야에서는 맡아서 작업하는 이의 의견을 최대한 따르고 존중해주는 방향으로 지도하였다. 또한 시간 조율의 측면에서는 항상 최악의 상황을 기반으로 스케쥴을 짜도록 지시하였으며, 실제로 많은 학생들이 처음 접해보는 3D 프린팅 프로세스에서 다른 팀들에 비해 시간을 절약할 수 있었고, 제 시간에 완성할 수 있었다. 이번 워크샵에서는 다양한 분야의 학생들이 새로운 문제를 해결하고 방향성을 제시하기 위해, 다각적 측면에서 문제를 바라보고 융합적 측면에서 문제점들을 해결해 나가는 값진 경험을 체험하는 동시에 3D 프린터와 멀티 미디어라는 새로운 미디어 방법론과 그 가능성을 얻어갔기에 의미 있는 경험이었으리라 생각된다.

KO Suejeong

With the topic of a Bio Digital City, it was our attempt to propose a new paradigm of future urban agriculture for Seoul where past histories and current diverse urban structures coexist. The architecture and urban knowledge of my team members helped to try the new method of 3D digital rendering and 3D printing as well as professional video works. It was a good opportunity to present the intention of these projects better through the collaboration of diverse methods and media.
Team B-1 selected Yeouido - the center of international affairs and finance which has a high density near the Han River Park. This team defines the Bio Digital city as a new city with algorithms of natural features implemented onto existing city structures. So they brought several natural features such as light and wind for urban agriculture to experiment and analyze using digital programs. Based on the results, they proposed a new urban format of the future optimized with an algorithm of land and changing variables.
Team B-2 chose Dangsan, the factory areas near Seonyudo Park and the Han River. This team suggested sustainable future urban agriculture derived from the international issue of the seeds war. It proposes to improve the outdated facilities of the site and connect it to Seonyudo Park to provide urban farming, leisure activities, and cultural experiences. Also, research based on the seeds bank exhibition and farming of the plants being studied was suggested to let citizens experience diverse sustainable urban agriculture of the future.

Bio Digital City라는 주제를 가지고 과거의 역사성과 현재의 다양한 도시구조와 특성을 가진 서울에 미래의 도시농업이라는 패러다임을 제안하고자 했다. 또한 팀원들이 가지고 있는 건축과 도시적 지식을 바탕으로 3D 디지털 렌더링과 프린팅 기술을 활용한 결과물과 전문가적 영상작업이라는 새로운 표현방법을 시도했다. 이러한 다양한 분야의 융합과 미디어를 통한 작품의 의도를 더 잘 표현을 할 수 있었던 좋은 기회였다고 생각한다.
B-1팀은 여의도라는 국제 업무와 금융의 중심지로서 한강 공원에 인접한 수변공간을 가지는 초고층 밀집지역을 사이트로 선정했다. 이 팀에서는 Bio Digital City의 개념을 기존도시에 자연요소를 도입한 알고리즘을 통한 새로운 개념의 도시를 만드는 것으로 정의한다. 그리고 여의도의 기존도시가로 체계를 바탕으로 빛이나 바람 등의 도시농업을 위한 자연요소를 도입하여 디지털 프로그램을 활용하여 실험하고, 분석한 결과를 토대로 땅이 가지는 알고리즘과 변화하는 요인에 최적화된 새로운 미래 도시형태를 제안했다.
B-2팀은 당산이라는 공장시설들이 모여있고 선유도 공원과 한강을 인접하는 사이트를 선정했다. 이 팀은 '종자전쟁'이라는 국제적 문제를 시작으로 도심에서 지속가능한 미래의 도시농업을 제안했다. 사이트의 낙후된 환경을 개선하고, 선유도 공원과 연계시켜 도심 속 농업과 레저 및 문화 생활을 제공한다. 또한 종자은행을 기반으로 연구하며 배양된 식물들을 전시하고 직접 재배할 수 있는 텃밭을 만들고 녹화된 벽을 활용하는 등 다양한 경험을 할 수 있으며 환경을 고려한 지속가능한 미래 도시농업의 모델을 제시했다.

KIM Dongjo

Convergence is one of the prominent social phenomena recently driven by the media industry. Our daily lives have been changing with new paradigms. With these changes, agriculture, the symbol of primary industry, is drawing attention as a future industry as it is combined with cutting edge technologies. With advanced technologies, agriculture in cities also plays a role as a green junction to connect eco-systems within urban areas not to mention its aesthetic function to improve the value of life in the cities.

This project produces creative outcomes with 3D printing by selecting potential sites in Seoul. Although it was a small scale prototype, students with architecture and media knowledge cooperated with other areas to reach their goal. It was an attempt to merge media and architecture to build the foundation of urban agriculture in the future. The concept was also based on the analysis of how natural features including water, wind and sun can define the main form of buildings not to mention the primary impact on plant's growth. The processes involve eco-friendly methodologies, materials and functions. The simulation of the buildings and outcomes were made into a video to show the potential to actualize the project.

최근 미디어 산업의 발전으로 두드러진 사회현상 중 융합을 꼽을 수 있다. 특히 새로운 패러다임의 형성으로 인간의 삶은 변화를 거듭하고 있다. 이러한 변화로 인해 1차 산업의 상징이었던 농업은 첨단기술과 융합되어 미래산업으로 주목 받고 있는 실정이다. 농업기술의 발전과 더불어 도시로 유입된 농업은 도시의 녹색 생태계를 연결해주는 거점으로서의 역할과 미적 경관 또한 도시의 가치를 향상시켜 준다.

본 프로젝트는 도시농업에 대한 연구를 기반으로 서울시에서 실현 가능성을 충분히 내재하고 있는 장소를 중심으로 재현될 건축시스템을 3D 프린팅 기술도입으로 창의적인 결과물을 제작하였다. 물론 소규모의 프로토타입 형태로 실행되기는 하였으나, 건축과 미디어 전공의 학생들을 중심으로 구성된 팀으로서 타 분야에 대한 이해와 논의를 거쳐 협업과정을 거쳤다. 이는 미디어와 건축의 융합에 대한 시도이며, 미래 도시농업의 근간에 대한 마련에 목적성을 둔다. 그리고 농업이 갖춘 자연친화적 환경요소들을 중점적으로 분석하여 건축물 제작에 대한 개념을 기획하였다. 자연요소인 물, 바람, 태양 등이 건축물의 형상를 결정하는 주요 요인이 되고, 식물의 성장과정에 대한 분석 내용들이 적용되었다. 이러한 제작과정은 친환경적인 공법 및 재료, 기능 등을 부합되게 만든다. 환경오염의 최소화를 가능하게 하며, 미래지향적인 창작과정을 경험할 수 있게 해준다. 그리고 실제 건축물의 시뮬레이션을 위한 영상제작으로 본 프로젝트의 발현 가능성을 시사하고 있다.

A TUFT CITY

BAE Jinsol JOUNG Taekwon KIM Dohee LEE Heeseung SHIN Yonghwan

Perspective 1

Appearance when viewed from the Han-River

Algorithm Of Nature

The concept of a Bio Digital City we suggest is to form new urban structures through algorithms of natural features such as light and wind. Respecting the existing buildings and roads of the city, the mass was designed through the shadows reflected at sunrise and sunset. We applied several computer programs (Vasari) to create an optimized space where humans and nature can merge in a harmonious way. Also, a city designed with those elements should keep the existing features of the current city to continue the contemporary flow, not to mention that it shows how people are longing for nature. This new urban structure has three benefits which are as follows:

우리가 생각하는 Bio Digital City의 개념은 기존 도시에 빛, 바람과 같은 자연의 요소를 적용한 알고리즘을 통해 새로운 도시체계를 구축하는 것이다. 기존 도시의 건물과 도로를 존중한다는 의미에서 해가 지고 뜰 때 투영되는 그림자를 통해 매스를 만들었고, 그 매스에 인간과 자연이 어우러질 수 있는 최적화된 공간을 만들기 위해, 실제 자연을 대입해 볼 수 있는 몇 가지 컴퓨터 프로그램(Vasari)을 적용시켜 인간이 실제로 살아갈 수 있도록 만들고자 한다. 또한 그 요소들을 적용시켜 만들어진 새로운 도시는 기존도시를 반영하여 그 흐름을 이어갈 뿐만 아니라 자연을 그대로 대입시켜 자연에 대한 인간의 동경을 나타낸다는 점에서 본질적인 의의를 가진다. 이렇게 기존도시의 흐름과 자연의 요소를 적용한 새로운 도시체계는 크게 세 가지의 이점을 가질 수 있다.

- It can build new urban structures that can adapt to the climate of the future while maintaining a reasonable amount of the existing city frame. Natural disasters are gradually increasing due to climate changes; therefore, a new architectural environment with data gathered from nature can be applied to protect humans from these disasters.
- An optimized urban agricultural system can be provided. Urban agriculture is a new alternative for a sustainable urban environment with expected future issues such as exhaustion of energy and food.
- To build an optimal city where human and nature can coexist with the intrinsic algorithm of each city. Each urban area has its own algorithm of land and environmental traits, so they will have diverse outcomes that will be differently optimized accordingly. These can be proposed as cities of the future.

-기존 도시체계를 어느 정도 따르면서 변화하는 미래의 기후환경에 적응하는 새로운 도시체계를 구축할 수 있다. 최근 변칙적인 기후로 인한 자연재해가 증가하고 있다. 자연의 데이터가 접목된 새로운 건축 환경은 인간이 자연의 재해 및 재앙으로부터 보호 받을 수 있을 것이다.
-최적화된 도시농업의 환경을 얻을 수 있다. 도시농업은 미래의 에너지의 고갈문제와 식량부족문제에 대응한 지속가능한 도시환경으로서 새로운 대안으로 제안가능하다. 자연의 요소를 적용한 도시체계는 자연에너지를 적극 활용되어 그 역할을 수행한다.
-각 도시가 가지고 있는 Algorithm에 따라 인간이 자연과 함께 살아갈 수 있는 최적의 도시를 만든다. 도시마다 그 땅이 가지고 있는 Algorithm과 환경적인 요인이 다르기 때문에, 그에 최적화된 각각 다른 형태의 도시가 만들어 질 것이고, 이를 하나의 미래형 도시 형태로서 제안할 수 있다.

Site

Yeouido residential districts located along the Han-River is at the heart of Seoul's financial and commercial hub. It is easily accessible and adjacent to the Han-River park

Perspective 2

Appearance when viewed within the city system

Eco - System

Systematize the symbiotic relationship between humans and plants

Digital might be perceived as a process that converts material into 0 and 1, and also as an opposite concept of Analog. However, Digital and Analog stand rather closely as extensions of each other. Likewise, the city with an algorithm of nature (bio) exists as the extension of the current city as it develops (digital).

The future of the Bio Digital City we imagined is both of memories and reflections of past images of its own. In order to remember the past, we reflected the existing city onto the new one and restructured the areas into optimized forms which then enables the coexistence of both agriculture and the human life style.

Bio Digital technologies in a Bio Digital City should be also properly selected: 'A TUFT CITY' in Yeouido is equipped with two:

- Considering the Han River in front of the site, we designed the growth of underwater creatures within the city; the residues from these creatures and water can then work as manure for other plants in the city and also as food for fish.
- We obtained the idea from the gills of fish for the external layer especially on the mechanism of how it opens and closes. When the gill-shaped layer opens at the bottom the wind from outside can come in, and when it opens at the top, it collects rainwater. Also, the facades are shaped in a diagonal direction which optimizes the collection of water.

Digital은 파생된 물질의 특성을 0과 1의 조합으로 바꾸는 과정으로 Analog에 대응하는 개념으로서 존재하는 듯하나, Digital은 Analog과 밀접한 관계의 연장선상에 존재한다. 마찬가지로 자연의 알고리즘(Bio)을 통한 도시는 본래의 도시를 잊지 않은 도시발전의 연장선상(Digital)에서 존재한다. 따라서 우리가 생각한 미래의 Bio Digital City는 과거 도시의 기억이자 투영된 거울이다. 과거의 도시를 기억시키기 위하여 기존 도시를 새로운 도시에 투영시켰고, 그 새로운 도시에 농업과 사람의 생활이 가능하도록 그 땅의 자연을 대입해 최적화된 도시로 재구성하였다.

Bio Digital City로서 Bio-Digital 기술도 도시에 적용시켜야 하는데, 도시에 따라 이용되는 환경이 달라지기 때문에 그에 알맞은 기술이 적용 될 것이다. 이번 여의도 'A TUFT CITY'에는 두 가지 Bio-digital 기술을 적용하였다.

-사이트 앞에 한강이 흐르고 있다는 것을 대입시켜 도시 내에서도 수생물을 기를 수 있도록 하였는데, 수생물이 남긴 찌꺼기와 물은 도시 내의 식물들이 잘 자랄 수 있는 거름이 되고, 식물에서 나오는 수분 속의 영양분은 물고기가 섭취하게 된다.

-물고기 아가미에서 고안한 외피인데, 물고기 아가미가 열고 닫히는 원리에 착안한 것이다. 아가미의 외피가 밑에서 열리면 외부의 바람을 실내에 유입, 외피가 위에서 열리면 우수를 집수할 수 있다. 또한 건물 파사드가 사선의 형태를 띠고 있어 우수를 집수하는데 유리하다.

Perspective 3

Appearance when viewed from the river

We utilized the light as a main natural feature in terms of creating new urban structures, we planned to activate urban agriculture on the most well-lit areas. The crops harvested through urban agriculture can be not only consumed but also used for commercial purposes similar to Highline Park in New York city which lends land to corporations both for farming and for public relations. The areas can also be used for sales related to the commercial facilities nearby. We also attempted to establish the urban agriculture system to both work as commercial and production advertising.
The dense mess at the bottom was the first issue to tackle; so we made a plaza there for people to walk through and then along the diagonal line, we then organized the mess to produce thin, consecutive planes to collect the light effectively in order to activate the system of urban agriculture. This was how we picture the three-dimensional structure of our design.

그리고 새로운 도시체계를 만듦에 있어 핵심적인 자연요소로서 빛을 이용하였는데, 빛을 가장 많이 받는 부분을 분석하여 현재 이슈화되고 있는 도시농업을 활성화하겠다는 계획을 가지고 있었다. 도시농업을 통해 수확한 농작물들을 단순한 식생용으로 사용하는 것뿐만 아니라 뉴욕 하이라인파크 사례와 같이 상업적인 용도로 이용하여 기업에게 농업을 할 수 있는 땅을 빌려주고 홍보할 수 있는 기능으로서의 역할을 하고, 상업시설과 연계하여 농작물들을 판매할 수 있는 기능도 도입하였다. 또한, 단순한 도시농업을 탈피해 상업과 생산 광고의 기능을 동시에 할 수 있는 획기적인 도시농업체계를 만들고자 노력하였다.
처음에 형태를 정리하는데 있어서 하부로 갈수록 밀집되어 있는 매스를 정리하여 사람이 거닐 수 있는 광장을 만들고, 빛에 따른 그림자의 형상에 따라 만들어진 사선의 매스를 정리하여 얇은 연속된 판을 만들고, 채광에 유리하게 만들어 원활한 도시농업이 이루어질 수 있는 도시체계를 스케치 투시도로써 생각하였다.

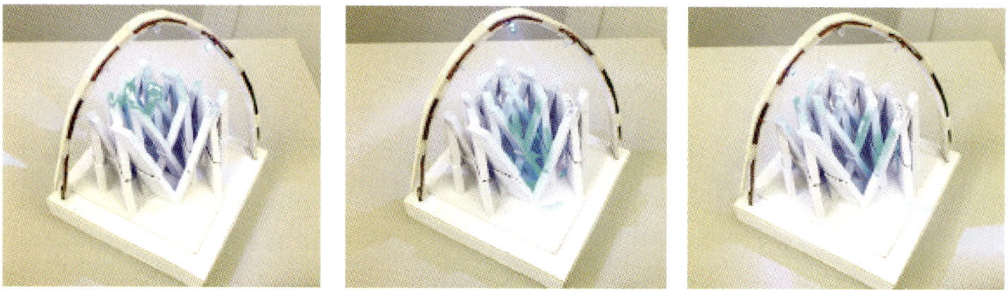

3D Model Printing

The initial design concept was to build new structures utilizing that of the existing city. So as a tool to approach this, we chose the light above all the various natural subjects. The processes of design were as follows:

- We made 4 planes by verticalizing the buildings and then analyzed the shadows over specific timelines to divide the areas of Yin(陰) and Yang(陽).
- We built a windpath based on surrounding natural aspects connecting proper programs to Yin and Yang areas in order to frame the optimization of urban agriculture and then placed support facilities such as residential, office and educational buildings.
- Focusing on the technology, we included a program to let human, plant, and underwater creatures coexist and designed the external layer with windows following the image of gills from fish.

Before producing a video, we tried to collaborate with visual art experts from the beginning. We wanted to let citizens participate in the workshop so we tried to make an introduction and design process video which would help people find the whole project more accessible.

- We put the process of framing out of the new urban structure using lights.
- With the Vasari program, we presented the simulation of outcomes in which we analyzed various natural features with lights.
- We presented each program on respective locations in our urban structure.
- We tried to show included systems for a technology perspective.
- We used a program called Lumion to let the audience experience the external and internal spaces of the design.

초기 설계 컨셉은 기존의 도시체계를 이용한 새로운 도시체계를 만드는 것이고, 그에 따른 도구로서 여러 가지 자연 소재 중 빛을 이용하였다. 그로 인한 설계과정은 다음과 같다.

-기존 도시체계 중 건물의 조직을 수직화시켜 4개의 면을 만들고, 빛을 이용하여 시간대 별로 생기는 그림자를 분석해 음과 양의 영역을 구분하였다.
-주변의 자연요소를 토대로 바람길을 만들어 그에 따른 형태를 다듬고 음과 양에 따라 생긴 영역을 각기 맞는 프로그램들과 결부지어서 빛을 많이 받는 부분에 도시농업을 하기 위한 최적의 형태를 도출시키고, 그 부분들을 뒷받침할 수 있는 주거영역 오피스 교육시설 등을 추가적으로 배치시켰다.
-기술적인 부분에 초점을 두어서 사람과 수생물 식물이 서로 공생할 수 있는 프로그램을 넣고, 건물 외피를 만듦에 있어 물고기의 아가미에 아이디어를 착안, 그 원리에 따라 창문을 만들어 외피를 결정지었다.

영상제작에 앞서 우리가 이번 프로젝트를 임할 때 단순히 건축만이 아닌 영상 전문가와 협력하여 결과물을 도출하고자 노력하였다. 시민들이 참여할 수 있는 워크숍이라는 주제에 맞게 모든 사람들이 친근하게 바라 볼 수 있는 인트로를 만들고, 그 다음 설계과정을 영상에 담기 위해 노력하였다.

-기존의 도시체계를 이용, 빛을 이용해 새로운 도시체계를 만드는 과정을 매끄럽게 영상에 담기 위해 노력하였다.
-빛을 통해 도출해낸 결과물을 바사리 프로그램을 이용하여 여러 자연 요건들을 통한 분석내용을 시뮬레이션으로 보여주고자 하였다.
-우리가 만들어 낸 도시체계에 각각의 프로그램들이 어디에 위치하고 있는지를 보여주고자 하였다.
-기술적인 부분에 있어서 우리가 도입한 시스템들을 보여주고자 하였다.
-루미온이란 프로그램을 통해 도시체계의 외부와 내부 공간에서의 느낌들을 투시도와 번갈아가며 체험하는 느낌으로 연출하고자 하였다.

Perspective 4

Appearance of the urban agriculture

It was our first time using a 3D printer at the workshop, we found it to be an ideal device to effectively express the initial concept though it had some limitations to represent the entire design. We encountered several problems. First, was the material restriction. While we wanted to maximize the penetrability of light by using transparent materials, we were unable to use the proper materials to achieve this. Second, 11 teams had to complete their printing within a limited time, but each printing took a lot of time not to mention that the printer itself is quite small which limits the size of the product. Lastly, it was printed gradually from the bottom up so there were a lot of issues in printing some unstable structures (in most cases, there were empty spaces in the middle). Once error occurred, the whole printing process stoped and the team had to start all over again - which then required detailed review before the printing could continue. Yet with improvement on these restrictions over time and after it becomes rather common, we believe it will help improve the design process a great deal especially in terms of saving time not to mention the expression of the products can be more detailed and completed in various ways.

이번 워크숍을 통해서 3D 프린트란 기기를 처음 접해보았는데, 아직까지는 설계 전반적인 내용을 표현하기에는 다소 무리가 있었으나, 초기 개념을 효과적으로 표현하기에는 안성맞춤인 기기였다. 다만 프린팅을 함에 있어 몇 가지 문제점이 있었다. 먼저 우리는 개념모형을 투명한 소재로 만들어 빛의 투과성을 극대화하려는 생각을 가지고 있었으나, 아직까지는 재료가 한정적이라는 단점이 있었다. 두 번째로 11개의 조가 한정적인 시간 안에 프린팅을 해야만 했는데, 프린팅을 함에 있어서 시간이 너무 오래 걸렸고, 기기자체가 작기 때문에 결과물이 크기가 다소 제한적이라는 점이다. 마지막으로 프린팅을 하는데 있어서 결과물 하부에서부터 단계별로 하는 시스템이기에 구조적으로 불안정한 결과물(중간에 비워진 공간이 있을 경우)일 경우에 하부에 구조체를 반드시 만들어 줘야 되기 때문에, 정형화되지 않은 결과물일 경우 프린팅 도중 오류가 종종 발생한다. 한번 오류가 발생하게 되면 프린팅이 이어지지 않아 처음부터 다시 해야해서, 결과물을 출력하기 전 세심한 검토가 요구된다. 하지만 시간이 지나면서 이러한 단점들이 보완되고 보편화 된다면, 앞으로 설계를 하는데 많은 시간을 절약할 수 있고, 결과물의 표현 역시 훨씬 더 섬세하고 다양한 범위로 나타날 것이다.

SELF-CONTROL CITY

KANG Huieun MIN Jihong PARK Kangeun

This site is in Seoul, located in the area of Youngdeungpo-Gu. This place is mostly industrial and includes a Lotte engineering factory, an auto repair factory, and a logistic warehouse. Dangshan is located between two bridges, past the purification plant which was the first ecology recycling park for environmental regeneration. The "water park" has a location advantage being adjacent to the edge of the water and is right in front of a national highway for the ease of transportation. While being close to the great features of the river and facing each other, this place is the center of a relatively stagnant area. Although there is residential development such as schools and other facilities for local factory workers, they are still rare and old which gives the area a more stagnant feeling. The resulting increases the risk of crime-ridden district. Thanks to the highway just across, the location is very convenient. However, in spite of the river being directly across the road, traffic blockage Create difficult accessibility to the park and the island.

본 사이트는 서울시 영등포구에 위치한 당산지역이다. 이곳은 대부분 공장시설로 이루어져 있고 현재 롯데 자동차 정비공장과 물류공장으로 사용된다. 당산부지는 양화대교 중간에 위치한 과거의 정수장 건축구조물을 재활용하여 국내 최초로 조성된 환경재생 생태공원이자 '물의 공원'인 선유도 공원과 인접하여 있다는 이점이 있다. 바로 앞의 IC 고속도로가 있어 전국으로 뻗어나가는 교통의 편리함과, 그리고 한강과 서로 마주보고 있다는 점도 큰 특징이다. 그러나 이곳은 서울의 중심이지만 주변에 비해 상대적으로 침체되어있다. 이유는 인근에 학교시설과 주거환경이 있지만 사이트는 공장지대여서 사람들의 인적이 드물고, 낡고 오래된 건물과 상대적으로 개발된 주변 건물들과 대조되어 더욱 침체된 면이 보인다. 이에 따른 우범지대의 위험성이 높아진다. 사이트 바로 맞은편에 IC고속도로 덕분에 교통에 매우 편리하다. 그러나 한강이 바로 앞에 있음에도 불구하고, 이 도로에 막혀 시민들이 한강공원으로 건너가려면, 다리를 이용해야 해서 사람들의 통행에 불편을 준다.

Concept Diagram

Phototropism / System Diagram

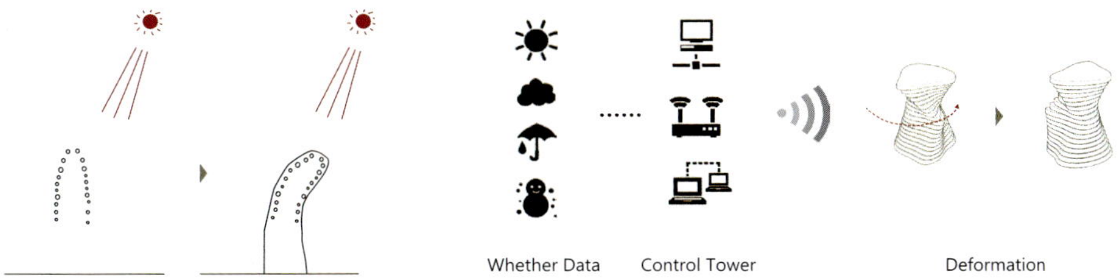

All of Earth's creatures have a reaction to a lot of stimulation from nature for survival. For example, in the case of plants, they have a certain mechanism which is called phototropism. There are several signaling molecules that help the plant determine where the light source is, and this activates several genes, which change the hormone gradients allowing the plant to grow towards the light. For a more efficient use of the environment, this project started from the idea that creatures have a responsibility for the change they create on their external environment.

This project gathers information from local climate data which is gathered over a long period of time from local climate data from the meteorology administration. This collected information can be brought to the control tower which is in the building. The control tower analyzes the information and sends a certain signal towards the skin of the building to deform it for more efficient control on the stimulation. As a correct example, imagine the control of the Heat-Island Effect. The Heat-Island Effect is a metropolitan area that is significantly warmer than its surrounding rural areas due to human activities. If the form of the building is changed in order to get fresh air and lower the temperature of the building according to the wind surrounding the building, the Heat-Island Effect can be controlled. In other words, proper modifications from the stimulation for more active control on the environment create a positive effect on the building.

모든 생명체는 개체의 생존을 위해서 자연으로부터의 자극에 대하여 일정한 반응을 한다. 이를테면, 식물의 경우, 태양으로부터 더 많은 빛을 얻기 위해 빛이 오는 방향을 향하는 굴광성을 보이곤 한다. 환경에 대한 보다 효율적인 이용을 위해서 이번 프로젝트는 생명체가 스스로 반응하는 것에서 착안하였다.

이 프로젝트에서는 장기간에 걸쳐서 누적된 기후 정보와 기상청으로부터 얻은 기상 정보를 수집한다. 여기서 수집된 정보는 건물을 제어하는 곳으로 보내지고, 여기서 분석된 정보를 기반으로 보다 높은 효율의 환경제어를 위해서 건물의 외형에 변화를 준다. 환경제어의 적절한 예로서, 건물의 내부의 열부하로 인한 열섬효과의 경우, 풍속의 변화에 따라 보다 많은 환기를 위해서 건물의 외형을 조절할 수 있다면, 적절히 제어될 수 있을 것이다. 따라서, 환경의 보다 적극적인 제어를 위해서 건물의 형태를 누적된 정보를 기반으로 변화시킨다면, 긍정적인 효과를 기대할 수 있을 것으로 생각된다.

Conceptual Sketch

Expection

As the current bridge, which connects Dangsan Area and Sunyudo-Park, is improved, the new bridge can attract not only local people, but also a lot of visitors who come here, with some new experiences such as farming in the city, and this will work to replenish the area. That is, putting in bicycle lanes from the Dansan Area to Sunyudo Park, planting a botanical garden which is based on the seed bank, creating urban farms for individuals in the city, the construction of a green wall, erecting a huge electronic display to guide people that gather there and to watch the big game or event. The new bridge can all be considered a landmarks in Seoul.
The building control system based on the data which is accumulated long time can modify the skin of the building and as it does, the building can provide good environment for the users. Also, the building control system can contribute to sustainable development by decreasing energy consumption and the emission of greenhouse gases such as carbon dioxide.
Before progress was made, various alternatives came out as the final plan. With the most basic cube type, an urban structure where each area has a different function such as residence, education, commerce, agriculture and green belt operate as a prototype and develops. Transportation with a motive of a Ferris wheel exists between each functional area. It was planned to make a society where self-sufficiency becomes possible through achieving production, transportation and the supply of crops.
The other alternative site was Ichon. The big flow of the other ALT starts with the program of the Seed Bank. The site of Ichon is the area where foreigners are concentrated and is close to Itaewon where a U.S. army base is located. In consideration of the characteristics of the site, the Seed Bank can keep and preserve the seeds of each country of the world and the most important seeds of our country, and in addition, the elements that form communities such as botanical gardens and farm-product markets that were derived from the program of the Seed Bank were thought of.
Combining these alternatives, Bio-Digital City where culture and leisure based on the Seed Bank can be enjoyed was applied to the undeveloped site that was previously used as a factory district called Dangsan and now, people often visit.

현재 당산지역과 선유도공원을 잇는 선유교를 개선함으로써, 도심 속에서 이루어질 수 있는 농업과 관련한 레저 및 문화 생활을 제공할 수 있고, 이 지역에 활기를 부여할 수 있을 것이다. 즉, 다리에 선유도공원까지 이어지는 자전거 전용도로와 종자은행을 기반으로 연구되고, 배양된 각종 식물 전시장소, 개개인을 위한 텃밭, 녹화된 벽과 경기 관람과 응원을 유도할 수 있는 대형 스크린을 설치함으로써, 다양한 경험을 할 수 있는 랜드마크적인 장소가 될 것이다.
장기간에 걸쳐 축적된 데이터를 바탕으로 외부환경에 따라 변화할 수 있는 건물제어시스템은 건물 사용자들에게 최적의 환경을 제공할 수 있고, 에너지의 사용량 및 온실가스 배출량을 줄일 수 있게 함으로써, 지속가능한 발전을 가능하게 한다.
최종 안으로 진행되기 이전에 여러 대안들이 있었다. 가장 기본적인 큐브의 형태를 가지고 주거, 교육, 상업, 농업, 녹지 등과 같은 각기 다른 기능을 가진 영역이 프로토타입으로 작용하여 증식하는 도시 구조를 만들었다. 각각의 기능의 영역들 사이에는 관람차를 모티브로 한 교통수단이 존재한다. 작물의 생산, 운반, 공급이 모두 이루어짐으로써 자급자족이 가능한 사회를 만들어보고자 하였다.
다른 대안의 사이트는 이촌이었다. 이 제안의 큰 흐름은 종자은행이라는 프로그램으로부터 출발했다. 이촌이라는 사이트는 외국인들의 밀집 주거지역이자 미군기지가 위치한 이태원과도 매우 인접한 곳이다. 이러한 사이트의 특성을 고려하여 세계 각국의 씨앗 종자와 가장 중요한 우리나라의 종자를 지키고 보관할 수 있는 종자은행을 넣어보려 하였고, 또한 종자은행이라는 프로그램에서 파생되어 식물원, 농산물 마켓 등과 같이 커뮤니티를 형성하고 사람들을 끌어들일 수 있는 요소들을 함께 생각했다.
이러한 대안들을 가지고 조합하여 생각해낸 것이 당산이라는 이전 공장지대로 사용되었던, 현재는 사용되지 않고 발길이 뜸해진 낙후된 사이트에 종자은행을 기반으로 한 문화, 레저를 즐길 수 있는 BIO-DIGITAL CITY를 구상했다.

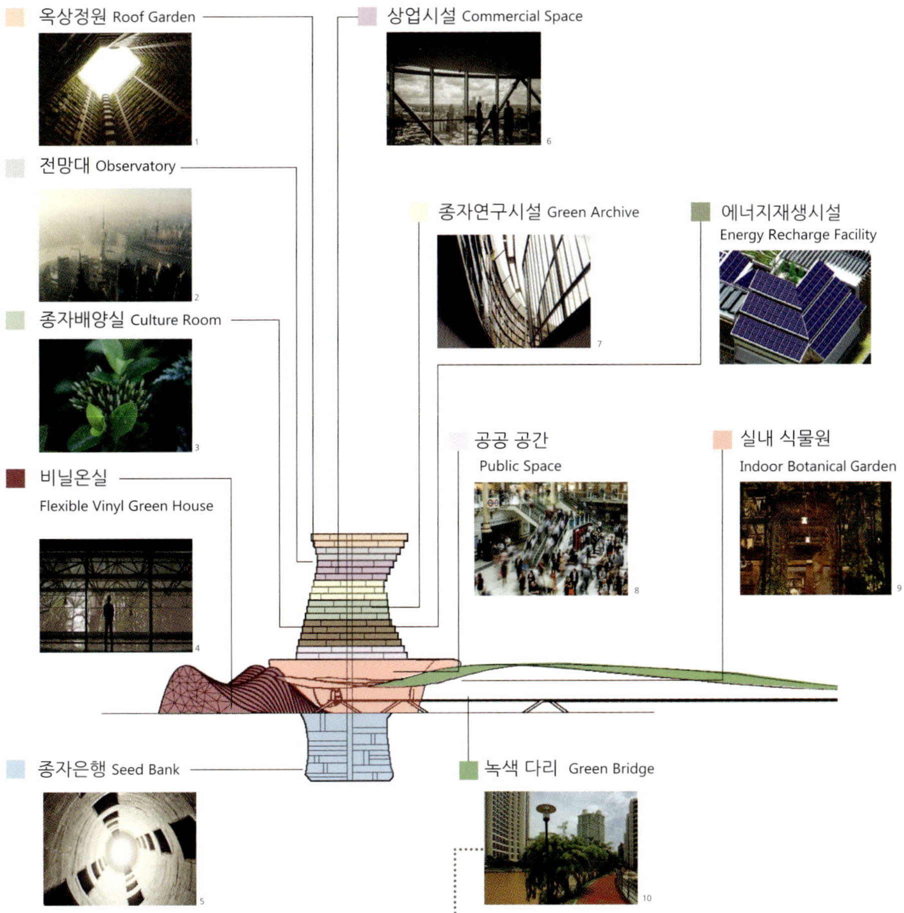

Program Section

Green Bridge

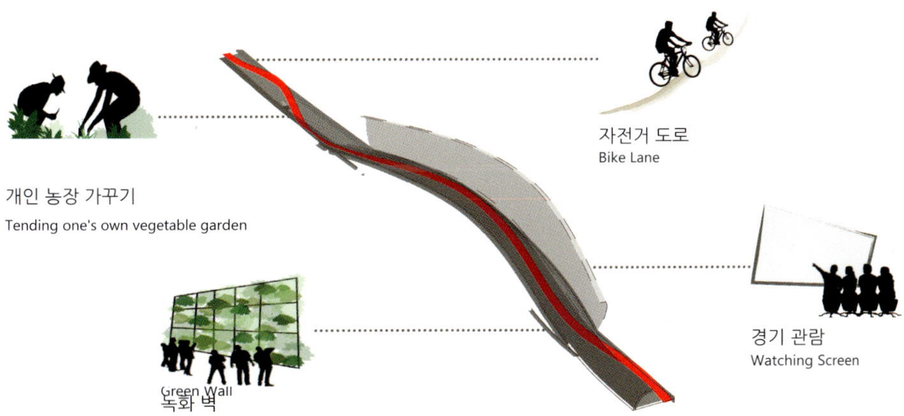

1. Redevelopment Catalysts

The seed bank represents an important opportunity to catalyze redevelopment adjacent to the current area of DangSan-Dong. A sustainable facility such as a seed bank can give a boost to the local area. Development typologies should be targeted for the neighborhood to create synergies between infrastructure, cultural amenities, and existing uses. Each development area should be established to be connected.

2. Creating an Infrastructure Connection

Infrastructure investments should be made within the public domain to facilitate connections between DangSan-Dong and the Han River. Infrastructure improvements should promote public access to key amenities within the cultural district (Seonyu Park). In addition, unique plants from the seed bank can be situated on the bridge and then this new botanical bridge can possibly become a landmark in Seoul.

3. Use of the Riverside

Through the green bridge between the seed bank in Dangsan-dong and Seonyudo-Park on the Han River, the connectivity with the Han-river can be improved. In addition to the inspiration created by the plants which is based on the seed bank, the use of riverside space can improve the quality of life in the local area.

4. Connectivity with Seonyudo Park

Seonyudo Park is located near the project site and should also be developed to have a connection to the rest of the area. Seonyudo (meaning "an island where Taoist hermits play") Park was the nation's first recycling park. It was created using the old sewage treatment facilities on Seonyudo Islet. Now, Seonyudo Park is a famous cultural space and there are various cultural programs that anyone can easily enjoy. The special experiences created from the seed bank and the green landmark can continue on into the cultural events of Seonyudo Park.

1. 재개발 촉매체

종자은행은 현재 당산동 인근 지역에 촉매제로서 중요한 기회를 제공할 수 있다. 종자은행과 같은 지속 가능한 시설은 지역사회에 활기를 불어 일으킬 수 있다. 개발지역은 주변 간접시설에서의 부수적인 효과를 일으키기 위해서 인근지역을 대상으로 해야 한다. 또한, 각각의 개발은 서로 연계성을 가지고 있어야 한다.

2. 구역간의 연결시설 확충

인프라 투자는 당산동과 한강 사이의 연결을 촉진하기 위해 공공 범위 내에서 해야한다. 인프라 개선은 문화 지구(선유도 공원) 내 주요 시설에 대한 공용 액세스를 촉진한다. 또한, 종자 은행으로부터 나온 식물은 다리에 위치할 수 있고, 이 새로운 식물 다리는 서울의 랜드 마크가 될 가능성이 있다.

3. 주변공간 활용

종자은행과 선유도공원을 연결하는 다리를 통해서 한강변과의 연계성을 증대시킬 수 있다. 종자은행에서의 식물이 주는 상쾌함과는 별개로 수변공간을 적극적으로 이용함으로써, 지역주민의 삶의 질을 증진시킬 수 있다.

4. 선유도 공원과의 연계

프로젝트의 사이트 인근에는 선유도공원이 위치하고 있다. 양화대교 중간에 위치한 선유도공원은 과거의 정수장 건축구조물을 재활용하여 국내 최초로 조성된 환경재생 생태공원이자 '물의 공원'이다. 지금은 서울의 대표적인 문화공간으로 각광받고 있으며 다양한 문화 프로그램이 준비되어 있다. 종자은행에서의 특별한 경험은 선유도 공원의 문화 컨텐츠로 이어질 수 있다.

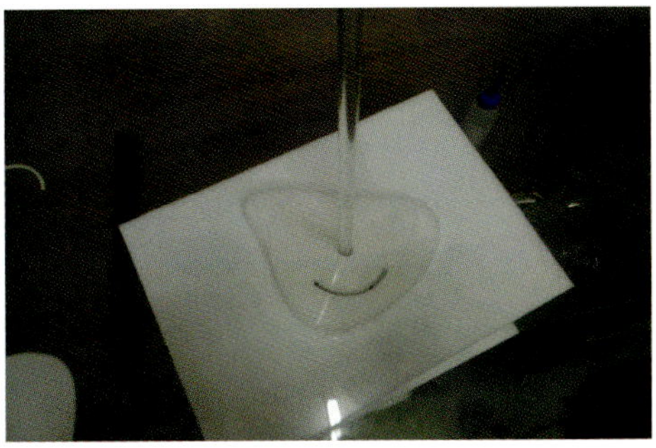

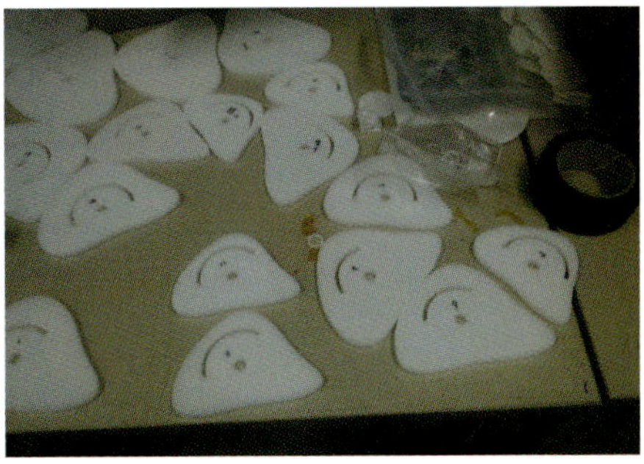

The video work started with the design of a storyboard. As for the scenario of the video, first of all the necessity of a Seed Bank and the seriousness in the current preservation of seeds are expressed with statistics and comics. Then, a scene containing the appearance of the whole city is shown after explaining the use of the program in each district, including the concept of buildings and bridges. Lastly, an imagined progress is contained when a residential area enabling self-sufficiency that is related in the future to urban agricultural facilities including only the Seed Bank program and its related culture, a leisure program is built.

The 3D printing work focused on the main building where the Seed Bank is located. Based on the concept that plants grow towards the direction of light, the program included in each layer was cut horizontally and can rotate according to the amount of light required. In order to express the movement of each layer, 3D printing progressed in the assembly method after the course of printing separately the same masses with boards of each layer. The masses of each layer have different shapes and sizes according to the programs but a natural flow of the whole building is made when it is rotating by preserving similarly the angles and shapes of masses as well as the connected top and bottom masses.

영상작업은 스토리보드 구상으로 시작되었다. 동영상의 이야기 흐름은 가장 먼저 종자은행의 필요성과 현재 종자 보존에 있어서의 심각성을 통계자료와 만화로 표현한다. 그리고는 건물의 개념과 브릿지를 포함한 각 동별 이용 프로그램을 설명한 후, 전체 도시의 모습을 보여주는 장면이 나온다. 그리고 마지막에, 종자은행 프로그램과 그와 관련된 문화, 레저 프로그램만을 담고 있는 도시 농업 시설에서 훗날에 이곳에 이와 연관된, 자급자족이 가능한 주거 지역이 들어섰을 경우를 상상한 진행과정을 담았다.

3d 프린팅 작업은 종자은행이 위치한 가장 주요 건물을 위주로 이루어졌다. 식물들이 햇빛의 방향에 따라 생장한다는 개념을 기반으로 수평으로 잘라진 각 층에 담긴 프로그램들은 햇빛을 필요로 하는 양에 따라 회전할 수 있다. 각 층의 움직임을 표현하고자 3d 프린팅은 각개층의 판과 같은 매스를 따로 프린팅하는 과정을 거쳐 조립하는 방식으로 진행되었다. 각 층의 매스는 담긴 프로그램에 따라 그 형태와 크기가 모두 다르나, 이어진 위 아래 매스와 매스의 각도와 형태를 유사하게 유지함으로써, 회전할 때 전체 건물의 자연스러운 흐름이 유도된다.

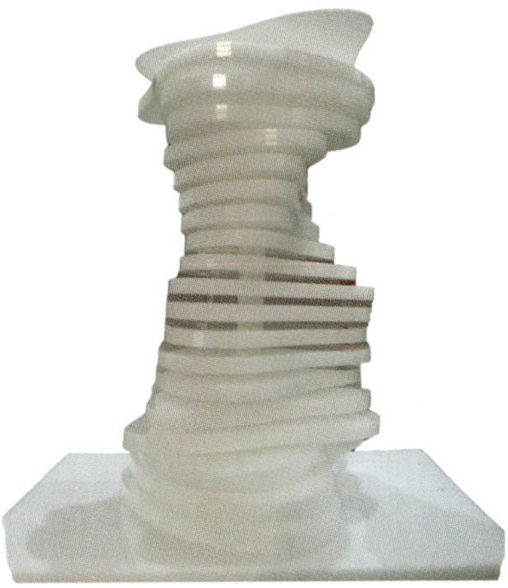

3D Printing Model

T - C

LEE Youngseok

How students perceived vertical farms was very diverse not to mention the intersecting processes to produce results based on their various understandings. The method to introduce vertical farming to the Hyundai Apartment complex, which is the most expensive area in Seoul, created opportunities for a lot of discussions and thought between the mentor and the students.

The development of cities brought to us many benefits but it also ended up physically dividing the core function of city and agriculture. Urban areas were developed without any agricultural features, ultimately they were created only for residential and commercial purposes. Sky-rocketing land value was one of the causes. So considering agriculture again in the urban context means many things to us. An increase in citizens' attentions was being drawn to the quality of life which naturally led to distrust of existing food especially processed food. Then it was connected to the realization that since we produce enormous amount of carbon dioxide this would be beneficial for crops that we could easily grow on our terrace instead of purchasing them in the stores. It also has an emotional side as well. Growing plants and consuming the produce from my terrace was once a delight of modern citizens, not to mention the joy you can get from the greens you yourself have produced. But this project doesn't include a discussion about whether urban agriculture is realistic or an ultimate solution for farming areas in cities and rural areas. Rather, it proposes experimental concepts of urban farming. So this project focused on the smartest way to adapt a new paradigm in the form of urban agriculture. I didn't want the students to focus too much on the buildings themselves but rather approach them from urban space perspectives. The smart plug-in of urban agriculture suggests that vertical farms can co-exist without any destruction to the existing apartment complex. Students from each team did a great job of following instructions and creating interesting results.

The green pots in drone form act in small units like living creatures which play their roles as a single unit and also as a collective to meet the essential dietary need of vegetables for one person. These vegetables are also delivered to the terrace of each house. A cloud-shaped structure was designed to take care of and control the crops located in parking lots. An agricultural unit in plug-in form for existing apartment was also considered. Everyone experienced restrictions to present their projects within the existing means. So they actively adopted media tools to focus on story-telling. The ideas were delivered better in a motion graphic presentation instead of presentation slides.

수직농장 혹은 도심농업에 대한 주제를 받아들이는 학생들의 시각은 매우 다양했고, 그 결과물을 만들어가는 과정 또한 흥미로웠다. 서울에서 가장 지가가 높은 주거지역인 압구정의 현대아파트단지 내에 수직농장을 소개하는 방식을 놓고 튜터와 학생은 많은 고민을 하게 되었다.

도시의 발전이 우리에게 많은 혜택을 가져다 주었지만, 이는 도시의 중심기능들과 농업을 공간적으로 분리하는 결과를 낳았다. 도심에는 농경지가 발 디딜 틈 없이 주거와 상업 중심으로 개발되었으며, 지가의 상승은 이러한 추세를 더욱 부추겼다. 도심에서 다시 농업을 고민하게 되는 것은 우리에게 여러 면에서 의미가 있다. 도시인들이 삶의 질에 관심을 가지게 되면서부터 먹거리에 대한 불신, 그리고 가공식품에 대한 반감이 커졌다. 간단히 베란다에서 재배할 수도 있는 작물들이 많이 존재함에도 이를 마트에서 사먹기 위해 매일 엄청난 양의 이산화탄소를 배출시키고 있다는 사실도 우리의 경각심을 일깨웠다. 정서적인 측면도 있다. 내가 키우고 가꿔 수확하여 먹는다는 것도 현대인들이 잃어버린 즐거움 중의 하나이다. 푸른 작물을 보며 얻는 시각적 즐거움 또한 크다. 그러나 이번 프로젝트에서는 도심농업이 도시와 농촌 또는 농업지역의 패러다임을 근본적으로 바꿀 수 있는 현실적 대안이 될 수 있을지 없을지는 논외로 한다. 오히려 실험적 의미로서의 도심농장을 제안한다. 이 프로젝트는 변해가는 시대의 새로운 패러다임을 수용할 수 있는 현명한 방식의 도심농업을 제안하고자 하였다. 학생들이 도심농업을 제안하기 위해서 도시공간적으로 접근하되 건축물 자체에 너무 집착하지 않았으면 하는 바램이 있었다. 기존의 아파트단지를 그대로 유지하면서도 수직농장을 공존시킬 수 있는, 스마트한 플러그인으로서의 도심농업이 되길 바랐다. 각 팀의 학생들은 그 가이드를 잘 따라주어 아주 흥미로운 결과물들을 도출해 내었다.

작은 유닛으로 나뉘어진 드론형태의 green pot들이 군집생활을 하는 생명체들처럼 모이고 흩어지며 그 기능을 다 한다. 유닛은 1인당 채소 필수 섭취량에 맞추어 구성되며, 이를 각 가정의 베란다로 운반도 가능하다. 주차장 대신 자리하게 되는 농장의 농작물들을 가꾸고 조정하는 도구로서의 구름 구조물을 고안하기도 했다. 기존 아파트 유닛 내에 플러그인 할 수 있는 농업유닛 또한 고민해 보았다. 모두가 기존의 프리젠테이션 방법으로는 자신들의 아이디어를 표현하는 것에 한계가 있음을 느꼈다. 스토리텔링에 더 많은 노력을 기울여야 하는 만큼, 학생들은 미디어 툴을 적극적으로 활용하였다. 슬라이드기반의 프리젠테이션이 아닌 모션그래픽기반의 프리젠테이션이 됨으로 인해 각 조가 생각하는 아이디어가 좀 더 완결성 있게 전달될 수 있었다.

KIM Seokwon

Though urban farming is by no means a new concept, it is recently attracting more attention with the rise of eco-friendly ideas and the notion of a compact city, the main issue of architecture. The goal of Studio C in this workshop was how to apply such concepts to an existing city. Especially to apartment complexes which have become the typical housing type in Seoul. An apartment complex was chosen as our site in order to create a program which can be applied to real and authentic conditions. We tried to make the most of the given conditions by minimizing demolition of existing buildings and the addition of new facilities.

In view of the limited schedule of the workshop, most of the work to develop the concept was carried out by digital modelling and the result was produced in the form of a movie clip and a model made by a 3D printer.

도시농업은 결코 새로운 개념이 아니지만, 친환경과 컴팩트 시티 등이 도시설계의 주요화두가 되고 있는 요즘 더욱 주목을 받고 있다. C조는 이러한 개념을 기존도시에 어떻게 접목시킬 것인가를 이번 워크샵에서 주안점으로 삼았다. 특히 그 대상지를 서울의 대표적인 주거형태로 자리잡은 아파트 단지로 설정함으로써, 현실적인 상황에 적용할 수 있는 프로그램을 만들어보고자 했다. 기존의 건물을 철거하거나 새로운 시설을 증축하는 것을 최소한으로 줄이고, 주어진 조건을 최대한으로 활용하는 것을 목표로 했다.

이 같은 개념을 제한된 워크샵 기간 내에 발전시키기 위해서 대부분의 작업은 디지털 모델링을 활용하였고, 결과물은 영상과 3D 프린터로 제작한 모델로 제작하였다.

SONG Sun

In modern society, people are exposed to media through various mediums but they seem to take the media as an image in itself. Media is not just an image which is expressed in one dimension but a means to send and receive 'messages' desired to be communicated through various mediums in our society. Now that we live in a society that communicates via messages in digital media, many fields are incorporating media as part of their strategy.

The 2014 Bio Digital City Tomorrow's Seoul Project was an opportunity for students with creative point of views to incorporate digital media into architecture.

The students who participated in the Bio Digital City Workshop were mostly students majoring in Architecture. When they received the theme of 'Bio Digital City', they didn't have any idea about the convergence of digital media and architecture because they also thought that digital media is no more than a series of images. This workshop was carried out as a tutorial for students to break the mold of their previous points of view and to think more creatively. The results were future-oriented, green city models.

Different fields are converging in modern society. Making students practice combining digital media and architecture, the Bio Digital City Tomorrow's Seoul Project did something meaningful in a sense that it helped students to synthesize other fields in the future.

현대 사회 속에서 우리는 다양한 매개체를 통하여 미디어를 접하고 있지만, 미디어를 단순히 영상매체라고 인식하는 경우가 있는 것 같다. 미디어란 일차원적으로 표현되는 영상매체가 아닌, 우리 사회에서 다양한 매체를 통하여 전달하고자 하는 '메시지'를 서로 주고받는 수단으로 여겨진다. 디지털 매체를 통해 메시지 전달을 하여 소통을 하는 사회가 되었기 때문에 많은 분야에서 미디어와의 접목이 이루어지고 있다.

이번 2014 Bio Digital City, Tomorrow's Seoul Project 에선 디지털 미디어와 건축의 융합을 학생들의 다채로운 시각을 통해 Bio Digital City만의 참신한 결과물을 도출할 수 있었다. Bio Digital City Workshop에 참여 학생들은 주로 건축을 전공한 학생들로 'Bio Digital City' 주제를 받았을 때 미디어와 건축의 융합에 대해서 막막해 하였다. 학생들 역시 디지털 미디어를 단순히 영상매체라고 생각한 부분이 있었기 때문에, 디지털 미디어를 생각하는 범위가 한정적인 부분이 있었기 때문에, 이번 Workshop에서는 학생들의 한정적인 생각의 범위를 넓히기 위해 기존에 생각했던 틀을 깨뜨리고 창의적인 생각을 할 수 있도록 많은 시간을 아이디어 도출과 생각을 다듬어 가는 tutorial로 진행되었다. 그 결과로 학생들의 참신한 아이디어들의 도출을 통해 자연과 더불어 살 수 있는 미래지향적인 도시 모습의 결과물들을 도출할 수 있었다.

현재 우리 사회에서는 한 분야만을 이야기하는 것이 아니라 타 분야와의 융합이 활발하게 이루어지고 있다고 생각한다. 따라서 이번 Bio Digital City, Tomorrow's Seoul Project는 학생들에게 디지털 미디어와 건축의 융합에 대한 고민을 하게 하였고, 추후 타 분야를 조화롭게 접목할 수 있는 계기를 마련해준 워크숍이었다.

G-CUBE HOUSING COMPLEX WITH VERTICAL FARM

PARK Jaehyun KIM Guntae KWON Chorok

Network Farming

Apgujeong Apartments

Before the commencement of this project, we decided to make a hypothesis to emphasize the necessity of vertical farming which is as follows: "The side effects of GMO foods, which had been hidden, started to appear explosively and people became unable to trust what was on their tables. They didn't trust what was in the markets and didn't think the law could guarantee the safety of their meals. So they actively brought vertical farming into their residential areas to regain trust in their food with ingredients that they grew themselves." Residents in the Apgujeong area are usually senior citizens and financially stable. They don't necessarily have to leave their current residence and wait for a new apartment to be built. So people should start by changing the materials of their current residential surroundings and adding agricultural space rather than adding construction or rebuilding, so that they can gradually change their way of life, eventually 'replacing' the whole.

프로젝트를 진행하기에 앞서 우리는 수직농장 필요성을 부각시킬 수 있는 가설을 세우기로 했고, 이는 다음과 같다 "숨어있던 유전자 조작 식품의 부작용이 폭발하듯 나타나기 시작했고, 사람들은 자신들의 식탁을 불신하게 된다. 시장에 나온 식재료를 신뢰할 수 없고, 법이 그들의 식탁을 안전하게 지켜줄 수 있다는 확신도 없다. 이에 그들은 수직농장을 주거단지에 적극 도입하여 자신들이 직접 기른 먹을거리로 식탁에 대한 믿음을 되찾으려 한다." 압구정에 주거하는 사람들은 재정적으로 안정되어있고, 연령도 높다. 이들은 굳이 살던 곳을 떠나 아파트가 신축되도록 기다릴 이유가 없다. 따라서 기존 주거환경을 새로운 재료로 바꾸고, 농업공간을 넣어 인위적 파괴와 재건이 아닌 각자가 삶의 양식을 바꿔 종래에는 이것이 전체가 될 수 있도록 '치환'한다.

THE MODERN RETURN TO NATURE : TRADE

We can't expect magnificent scenery in an urban area, however, we can find a way to enjoy whatever suits the conditions and surroundings. With increasing anxiety about food safety after the harmful effects of GMO foods were revealed, if we make spaces to grow crops in houses and residental areas, we can experience the beauty of nature directly. The modern way to return to nature is not just about adding agricultural spaces. As capitalism increases, people judge the worth of everything by its monetary value. Therefore, the trade of crops will let people return to the past before capitalism and think rather deeply about the actual meaning of trade and not the monetary value.

OPERATION OF VERTICAL FARMING

Self sufficiency through farming is their goal but the variety of crops that can grow within individual houses is limited, so they need to be provided with crops they will not be able to grow themselves. So the goal is to plan for vertical farms within the residential community and place them near the main roads as a representation of the community. Also, these farms will be responsible for the crops that can't be planted in individual houses. Going further, selling guaranteed non-GMO seedlings will create a secured provision of trustworthy food for residents. The vertical farms will be run as follows: Residents will be entrusted with the crops for certain areas of land. All the costs for these entrusted crops will be covered from seeding to harvesting but money is also invested into the initial operation of the vertical farm. After a certain level of stabilization, money can also be invested for the development of the farm to the next level. Therefore, they can not only be provided with crops grown precisely for them but also any surplus production exceeding the needs for trade can be sold outside of their communities. This will generate profits which will then be distributed back to the residents. So it is not just a simple farm, it is comprised of proactive production activities by residents themselves as they are participating in the operation as well. This is how we are targeting to lead the new culture of urban residential areas with this system spreading from one community to others.

도시 내에서 금수강산의 풍광을 탐미하기를 바랄 수는 없다. 하지만 뭐든 상황에 맞게 즐길 수 있는 방법을 찾으면 된다. 유전자 조작식품의 폐해로 식품안전에 대한 불안감이 나날이 높아지는 상황에서 가내, 단지 내에서 작물을 재배할 수 있는 공간을 만들어 내 손으로 안전한 먹을거리를 가꾸어낸다면, 함께 자연을 간접적으로 체험할 수 있는 것이다. 소제의 현대적 자연으로의 회귀는 단순히 농업공간을 넣는 것만을 말하고자 하는 것은 아니다. 극도로 치닫는 자본주의 사회에서 살아가는 사람들은 돈의 절대적 가치에 의해 모든 것을 판단한다. 따라서 작물의 교환에서 만큼은 자본주의를 버리고 과거로 돌아가 물물교환을 통해 돈에 의한 가치 판단이 아닌 교환의 의미를 더 깊게 생각할 것이다.

농업을 통한 자급자족이 그들의 목표이지만, 가정에서 재배할 수 있는 작물의 범위가 한정적이므로, 그들이 재배하지 않는 작물을 공급받을 수 있어야 한다. 따라서 단지 내에 수직농장을 계획하고, 이를 대로변에 배치하여 단지의 성격을 드러내는 상징적인 요소로 삼고, 가내에서 키울 수 없어 별도의 공급이 필요한 작물들을 위탁 재배하는 것을 목적으로 한다. 또한, 수직농장에서 키워 유전자가 조작되지 않았다는 것을 보장할 수 있는 건강한 모종을 판매함으로써, 주민들이 믿고 먹을 수 있는 먹을거리의 공급에 힘쓸 수 있도록 한다. 수직농장의 운영방식은 다음과 같다: 주민들은 '주'에 해당하는 재배면적을 사 작물의 재배를 위탁한다. 이 '주'의 비용은 작물의 파종부터 수확까지 전 과정을 관리하는 것에도 사용되지만 갓 도입된 수직농장의 운영비용에도 사용된다. 어느 정도 안정화된다면, 개발 발전비용에 사용될 수도 있을 것이다. 이로써 이들은 자신들만을 위해 재배되는 작물을 얻을 수 있음은 물론, 물물교환의 수요를 초과한 잉여작물들은 단지차원에서 외부로 판매해 수익을 얻고, 이는 다시 주민에게 분배된다. 단순히 수직농장을 도입하는 것만이 아니라 입주민이 직접 운영에 참여하여 생산 행위를 주도한다. 우리는 이러한 시스템이 한 단지에서 도시로 확산되어 새로운 형식의 주거문화가 형성되는 것을 주도하고자 한다.

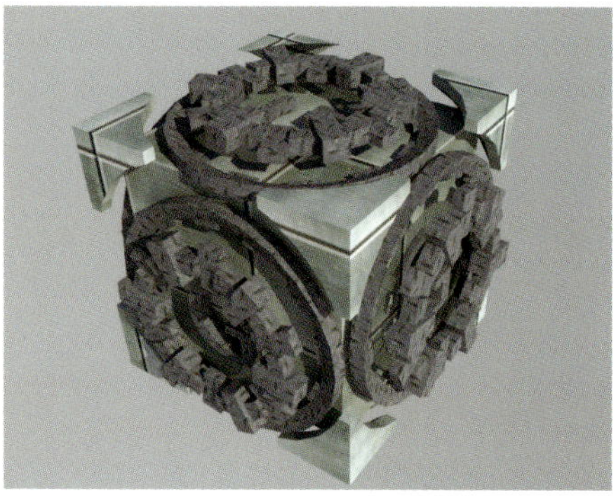 

G-Cube

Main concept

CARBON NANO TUBE

The residential space is not limited to only concrete walls but will be formed with a G-cube combination made with Carbon nano tubes. A carbon nano tube is a compound of a carbon molecule and three other carbon molecules forming a hexagon honey comb shape. It is known for being light-weight but also high in solidity and conductivity. Utilizing these traits, it can be used to build structurely stable housing that is stronger than ferroconcrete buildings and also improves the connection between households and communities. Furthermore, it is easily adhesive and running a current through elements that are sensitive to external devices which can then quantify the amount of carbon dioxide emitted by the living creatures to automatically control their size accordingly. Thus the size of the space is not fixed but the G-cube recognizes the changes of carbon dioxide emitted through respirations to decide the size of the space accordingly. Internal composition of the unit is also not pre-determined but flexible such as trading LOT. You can buy and sell LOT as much as necessary which enables the separation and extension of the space. It provides much more flexibility to the urban life style. We aimed to achieve the gradual replacement from small units to replace the whole system, as substituting the materials can lead to bigger changes within the frame of flat type apartments rather than destroying the entire apartment complex and re-building it.

주거공간은 더 이상 콘크리트의 벽체로 한정되는 것이 아닌 탄소나노튜브로 만든 G-cube의 결합으로 형성된다. 탄소나노튜브는 1개의 탄소 원자가 3개의 다른 탄소 원자와 결합한 육각형 벌집 모양의 속이 비어있는 튜브형태의 구조로 가볍지만, 강도가 높고, 전기전도율이 높다. 이러한 성질을 이용하여 철근콘크리트 이상의 강도를 가진 구조적으로 안정된 집을 만들 수 있고, 각 세대와 단지 간의 네트워크 성능을 강화할 수 있다, 또한, 무엇이든 잘 붙는 성질과 소자를 통해 흐르는 전류가 외부기체의 흡착 여부에 민감하게 반응한다는 성질을 이용하여 생명체가 이산화탄소를 배출하면, 그 양을 수치화하여 인식한 후 스스로 공간의 크기를 조절할 수 있도록 한다. 즉, 공간의 크기가 한정되어 있는 것이 아니라 호흡을 하며, 내뱉는 이산화탄소의 양의 차이를 G-cube가 인식하고, 그 양에 따라 공간의 크기가 결정되는 것이다. 이에 의해 유닛 내부구성도 자유롭게 변할 유닛의 형태 자체도 고정된 것이 아니라 LOT을 매매하는 것과 같이 변화할 수 있다. 내가 필요한 만큼의 lot을 자유롭게 사고팔며 공간의 분할과 확장이 가능해져 삶의 양식이 보다 유동적인 성격을 띨 수 있다. 우리는 이를 통해 기존 판상형 주거시설의 틀 안에서 건축물의 파괴와 재건이 아니라 재료의 치환이 점차 큰 단위의 치환으로 이어지며, 이렇게 변환된 것들이 곧 전체가 된다는 것이다.

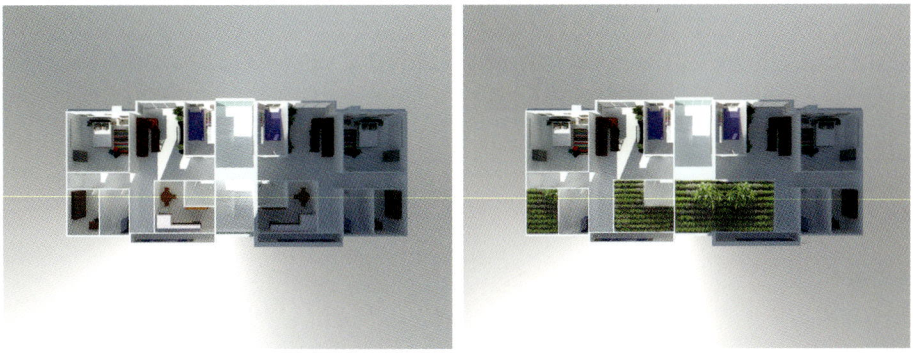

Single Unit Substitution

Changed by G-cube

Detail Of Changing Space

Media effect / Farming

Story Board

In order to manage trading, the new form of consumption, and to improve the relationship between efficient in-house farming and vertical farms, a media called 'Smart Hub(Herb)' is to be introduced. 'Hub', in general, means the accumulation point of multi-directional data flows. But we also added the symbolic homophone 'Herb' to it in order to represent our intentions better. It provides data about the in-house crops and also how the entrusted crops are growing. After harvesting, each house can enter in the surplus amount of the product to inform others so that they can reserve the crops they need. This will help promote the trading markets. Also, the open market trade prices on the system will promote the practice of fair trade. Therefore it not only guarantees the provision of safe foods but also covers the whole process from growing the crops to distribution.

가내 작물의 효율적인 재배와 수직농장과 가정간의 관계성을 높이고, 새로운 소비의 개념인 물물교환을 관리하기 위해 '스마트 허브'라는 미디어를 도입한다. 통상적으로 사용되는 스마트허브의 '허브'(Hub)는 하나 이상의 방향에서 데이터가 드나드는 데이터의 집적장소를 말한다. 하지만 우리는 여기에 작물을 상징적으로 비유한 'Herb'의 의미를 넣어 그 목적을 보다 잘 표현할 수 있도록 했다. 이는 집에서 재배하는 작물에 대한 정보를 알려주고, 위탁한 작물의 재배가 어떻게 이루어지고 있는지를 볼 수 있다. 수확한 후에는 각 집에서 잉여량을 입력하고 이를 공개하여 자신이 필요로 하는 작물에 예약을 거는 등 물물교환시장의 교류가 활발하게 이루어 질 수 있도록 도움을 줄 것으로 예상한다. 또한, 이 시스템에서 제공하는 물물교환 시가를 통해 공정한 거래를 할 수 있다. 즉, 이것은 안전이 보장된 식품을 섭취할 수 있도록 함과 동시에 작물의 성장에서 유통에 이르는 과정을 총괄하는 역할을 한다.

In the bigger picture, we produced a 3D model of a flat type apartment frame and the internal composition of the unit within. This aimed to present how a small replacement in the internal environment, like new materials, can gradually spread toward bigger changes. It is a steady process of expanding from Nano sized material replacements towards an apartment and then to a whole complex or residential community.
A delay that happened at the beginning affected the whole project more than expected. We had numerous meetings beforehand, but the concept idea didn't reach a level that satisfied every member, which impacted our team until the beginning of the workshop. We went through hard times but we supported each other and believed that something good would come up in time - and finally we came up with a great concept idea.
Based on the concept of replacement, we wanted to show how certain things within apartment units could be replaced - like drawers. But the overall progress was delayed so we didn't have enough time for 3D modeling. Considering all the obstacles, it was going relatively smooth until the deadline was just around corner. The first issue was that we missed an error with the bonding method in a sketch-up program since it was our first time trying 3D printing. We constantly got an error message that printing couldn't proceed - this costed us several trials. The second issue we had was that we had too many units to print within the given time limit. Actually, the products that were printed out arrived a day before the presentation and were much less than what we expected. We planned for the same amount of units as the actual Hyundai Apartment which should have been one flat type apartment frame and 48 units. It required more than 100 hours of printing, and that was something that we could not do given the time restrictions. The lack of units made our model look awkward with all the empty spaces and we thought it might have failed to show what we initially intended. Now looking back at that experience, we should have printed a G-cube instead to show the bonding method to better explain the concept of our design and overall ideas. We were all new to 3D printing and the lack of understanding also affected the outcome. We should have done more thorough research to figure out the most optimized way to present our project.

우리가 뽑아낸 3D 모델은 크게 보면 판상형 아파트라는 틀과 그 내부를 구성하는 유닛을 출력한 것이라고 할 수 있다. 이는 G-cube라는 새로운 재료의 도입으로부터 시작된 내부공간의 변화가 점차 큰 범위로 치환되는 것을 나타내고자 함이었다. 오랜 회의 끝에 마침내 좋은 컨셉 아이디어를 얻게 되었다. 치환이라는 개념을 본바탕으로 서랍장처럼 아파트유닛을 여러 가지 유닛으로 치환할수 있다는 것을 보여주고 싶었다. 하지만 전반적인 작업 속도가 늦어져 3D 모델링을 위한 시간이 상대적으로 부족해졌다. 이런 상황에 비해 3D 프린팅 작업은 비교적 순조롭게 지나가는 것 같았다. 하지만 마감을 코앞에 두고 문제가 발생했다. 첫 번째 난관은 3D 프린팅을 처음 접해봤기 때문에 스케치업 프로그램에서 결합방식의 오류를 인지하지 못했던 것이다. 계속해서 프린팅을 할수 없다는 오류메세지를 받게되었고 몇 번의 시행착오 끝에 프린팅을 할 수 있게되었다. 두 번째 난관은 출력할 수 있는 시간에 비해 유닛의 양이 너무 많다는 것이었다. 실제로 최종 발표 전날 도착한 출력물에는 우리가 생각했던 것보다 훨씬 적은 양의 유닛이 들어있었다. 실제 현대아파트의 유닛개수와 같게 했기 때문에, 판상형 아파트의 외장틀과 48개의 유닛이 필요하였다. 100시간이상의 프린팅 작업이 필요했고, 우리에겐 그만큼의 시간이 없었다. 부족한 유닛으로 인해 빈 틀 안에 군데군데 들어가 있는 유닛들은 허전한 느낌을 주었고, 우리가 의도했던 바가 무엇인지도 잘 보이지 않았다. 시간이 지나고 나서 생각해보니, 오히려 G-cube를 출력하여 그 결합방식을 보여주는 것이 설계 개념은 물론 전반적인 주제를 표현하는데 더 적합하지 않았을까하는 아쉬움이 남는다. 더욱이 3D 프린팅 작업은 모두 처음 접해보는 것이었기 때문에, 다소 생소해 그 특성에 대한 이해가 부족했던 것도 한 몫 했던 것 같다. 3D 프린터에 대한 조사를 보다 철저히 하여 더 효율적으로 표현할 수 있는 방법을 찾았어야 했다.

3D printed model

B-POT PROJECT

AN Seongseop KIM Minwoo KIM Hayan

Center for B-POT

Where can humans live most humanly? During Korea's modernization, a lot of the population concentrated in Seoul, and it became full of people, cars, concrete, and polluted air. The high mountains and fields of Hanyang (the old name of Seoul) were lost and only small and shabby bushes with cobwebs and trash remained. The small pleasure of growing their own plants, in their own places of residence, started from the Stone Age when mankind started to collect their food from the fields and forests to survive. The ages continued on and the fantasy about the ultimate goal of human life has been to live in tune with Mother Nature. It doesn't matter how fancy your house is or how much money you possess or even the quality of the car you own. What matters most is not being concerned with breathing fresh air in a confined space. Throughout the industrial revolution, the separation of city and country began to be much more apparent, the number of people who do not benefit from fresh air and the presence of nature has grown exponentially . Accordingly, those who live in the city are trying to attract any kind of nature near to where they live. They start by planting their crops in small quantities in small pots put at the windows to make their living space into a tiny farm in the city, which sometimes provides a farm experience for these citizens. As humans have evolved and adapted to the urban system, farming also starts to change through the development of the city. How can green spaces and farms for urban people develop much more than modern farms and how can these things become harmonious within the city's system which is full of digital media programs? We will start with our proposal about a green bio digital city for Seoul.

사람이 사람답게 살 수 있는 공간은 어디일까? 한국이 근대화되는 동안 수많은 인구가 서울로 집중되었고, 그만큼 서울은 사람과 콘크리트 오염된 공기로 포화상태가 되었다. 한양을 채우고 있던 드높은 산과 들은 이제 다 사라지고 서울엔 차량 옆 거미줄이 쳐진 초라하고 생명력 질긴 생물들만 남게 되었다. 나의 공간에서 소소하게 내가 원하는 식물이 자라는 것을 보는 즐거움은 인류가 채집으로 식량을 구해 먹던 구석기 시절부터 시작되었고, 시대가 지나면서도 인간의 궁극적인 좋은 삶에 대한 환상은 항상 좋은 자연 옆에서 생활하는 것이었다. 좋은 삶의 질에 대한 불변하는 법칙은 항상 자연과 함께하는 것이다. 아무리 좋은 집과 많은 돈과 차가 있더라도 한정된 공간 안에 맑은 공기를 마시며 살 수 없다는 것은 결코 좋은 삶이 아니다. 산업혁명을 거치면서 도시와 지방의 분리는 더욱 확연해졌고, 맑은 공기와 자연을 누릴 수 없는 사람들의 수는 늘어만 갔다. 그에 따라 도시에 사는 사람들은 자신이 사는 곳 근처에 조그맣게라도 자연을 끌어들이려고 노력한다. 작은 화분에 원하는 작물을 소량으로 심어 창문에 두는 것으로 시작하여 도시 속에 농장을 만들어 도시민들을 위한 농장체험을 제공해주는 경우도 있다. 인간이 도시에 적응하듯 농장도 도시에 적응하며 발전해갔다. 과연 도시주거민들을 위한 녹지공간과 농장은 더 어떻게 발전할 수 있을 것이며, 도시의 디지털 미디어적 시스템과 어떻게 조화를 이룰 수 있을 것인가? 이번 프로젝트의 제안을 시작하려 한다.

BACKGROUND

The proportion of Seoul's citizens who have experienced city farming is about one-third of the total population, specifically 34.5% of Seoul's citizens have seen ongoing urban agriculture. The first reason for the popularity of this type of urban agriculture is that the citizen is involved, showed by the highest rate of 66.1%, that they believed the safest food to eat was that of their own cultivation. Secondly, health and leisure were important as well. It was found that people take only a short period of time to do their urban farming.

The greatest barriers to urban agriculture is lack of space.(1) Based on this, we propose a system that an individual can easily access, ensures safety from agricultural chemicals, and uses space rationally for agriculture. The basis for the proposal started from concerns about what is seen as the biggest issue of Seoul's housing problem - is there room to grow crops when there is no space for people to reside in? The solution is not to use ground space but instead the empty space above. This method is the most efficient and convenient, but technology and machinery is required in order to use this space above. That's why the B-POT project has been designed. Over time, many people will design systems set by their own movement, needed in order to live conveniently. In addition, this will be a primary system in the future because of its multi-functionality which acts as a role of one or more objects. Therefore, the future system of farming also should be multi-functioning. The B-POT service, according to the characteristics of contemporary people's favor, offers a small garden and if there is not enough light for the develop, they will function as media wall.

(1) reference : Present condition and development of urban agriculture of Seoul / Kwon Taeeun / 2014.01

서울 시민 중 도시농업을 겪은 사람의 비율은 전체의 34.5%정도로 3분의 1정도의 사람들이 도시농업을 겪어봤다. 또한 도시농업을 하는 이유는 농작물을 안전한 먹거리를 먹기 위해가 66.1%로 가장 높은 비율을 보였으며, 건강과 여가가 그 다음을 이었다. 또한 농사를 짧은 기간 동안 짓는 것을 알 수 있었다. 도시농업의 장애요인으로서는 공간부족이 가장 큰 이유가 되었다.(1) 이를 바탕으로 개인이 손쉽게 접할 수 있으며 안전을 보장받는 농작물재배와 공간에 대한 고민 없이 쉽게 접할 수 있는 도시농업의 시스템을 서울시에 제안한다. 제안에 대한 기본은 서울시의 가장 큰 문제 중 하나인 주택문제를 보듯이 사람이 살 공간도 없는데 과연 농작물을 심을 공간이 있을 것인가에 대한 고민부터 출발한다. 땅에 디디지 않고 공간을 쓸 수 있는 방법은 공중을 사용하는 방법이다. 가장 효율적이고 편리하지만, 공중공간을 사용하기 위해선 기술과 기계가 필요하다. 이에 따라 프로젝트의 중심인 B-POT이 설계되었다. 시간이 지날수록 많은 사람들은 살아가는 데 필요한 시스템을 자신의 움직임에 맞추어 편리하게 이용하도록 설계할 것이다. 또한 한가지의 목적만을 위한 물건을 만들기보다 한가지의 역할을 하면서 또 다른 역할을 할 수 있는 멀티기능성을 지향하기 때문에, 미래의 시스템은 다양한 기능을 수행할 수 있어야 한다. B-POT은 이러한 멀티기능성을 지향하는 현대인의 특성에 맞춰 개인 소규모 농장을 제공하며, 빛이 없어 효율성에 따라 움직일 수 없을 땐 디지털 미디어 월의 역할을 수행한다.

Evolution of B-POTS accending to sun location

Morning / Afternoon / Late afternoon / Night

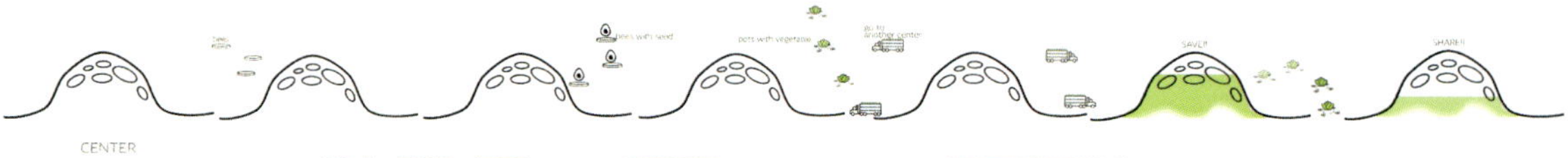

B-POTS at the veranda

They move on apartment's veranda when people want to see their crop.

PROPOSAL

To solve the fundamental problem of space, we need to find a technology that optimizes the use of the areas above the ground that is both free and accessible for individuals at anytime and in any place. The concern about the space is solved by light efficiency and people's sight. The middle of the apartment is useful because there is sufficient light, and people in the apartment can easily see the plants in the pots. This project also intends to actively use the balconies, so B-POT will be hung on the balcony's handrail and will be controlled by the owner.
The B-POT needs two parts of machinery. The first part is the fixed center of the system where the plants grow. The second part is composed of the moving parts which move the POT to a place where it can get the most sunlight. The moving of a big POT is too dangerous and inefficient, so a little Bee follows where the moving activity is needed and acts as a media wall. Therefore, there are no fixed features that contemporary architecture usually has, the form is built by the shape of the moving B-POTs which move to spaces that have sufficient sunlight and that the owner can easily approach. The shape is constantly changing and that is its architecture.
The central control tower determines the location of the day's sunshine and clouds to the B-POT, and controls them to move slowly to a specific location. When the dark of night approaches, the Bees start to illuminate themselves and act as the media wall of the apartment or as a street lamp for people who are getting off from work late. The media wall will be the identity of the apartment.

근본적인 공간문제를 해결하기 위해 공중공간을 활용할 수 있는 기술과 더불어 개인이 언제든 편리하게 접근 가능하도록 만드는 기술이 필요했다. 아파트 단지의 어느 공중부분을 사용할 것인가에 대한 고민은 B-POT들이 쉽게 햇빛을 받을 수 있는 부분과 주민들이 볼 수 있는 곳, 또한 가까운 거리에 위치해야 한다는 점을 고려하여 동과 동 사이에 있는 공중공간을 사용하기로 하였다. 또한 아파트 동이 가지는 베란다 공간을 활성화하여 식물을 기르는 고정적인 주 부분이 필요하고, 그를 옮기면서 연결고리 역할을 하는 부분 총 두 가지가 필요하다. 개인의 소규모 농장이 스스로 움직이는 것은 너무 크고 불필요한 움직임을 많이 포함하기 때문에, 식물이 움직이지 않아도 되는 부분은 작은 부분(BEE라고 일컫는다)이 처리하고 개인이 직접 농작물을 보고 만지고 기르고 싶을 때는 개인소규모농장(B-POT이라고 일컫는다)을 BEE가 옮기는 형식으로 형태를 구성한다. 따라서 지금의 건축이 가장 보편적으로 가지고 있는 고정적인 형태는 있지 않으며, 이 B-POT들이 자신의 필요성에 의해 햇빛이 많은 곳으로 혹은 사용자가 접근하기 쉬운 방법으로 이동하면서 형성되는 모습 자체가 건축물이 되고 구조가 된다. 중앙 컨트롤 타워에서 그날의 일조량과 구름의 위치를 파악, B-POT들이 햇빛을 가장 효율적으로 받을 수 있는 장소로 서서히 이동하게 되며, 밤이 될 경우 밤길을 밝히거나 한강변의 똑 같은 아파트들 사이에서 독특한 아이덴티티를 가지고 집을 홍보할 수 있는 미디어 아트 월의 역할을 하게 된다.

EXPLANATION OF B+POT

There are three main parts to the B-POT. There is the plant pocket where the plant can grow and the holes in the BEE which can move the POT. The wing part that the BEE holes are a part of can be folded easily to form a variety of shapes. This flexible architecture gives sustained support to each one of the connected POTs. The BEE holes where the Bee can attach are designed to be well equipped with an upper side that is narrow and a bottom side that is wide so the Bees are unable to escape. B-POTS are moved by collective intelligence. Development of ICT (Information & Communication Technology) technology and IOT (Internet of Things) technology makes it possible for modules with intelligence to know where they are needed and where they have to go. They can control themselves. One B-POT may provide vegetables for two people for one week and each home is subjected to actual sunlight instead of the artificial light in a room, so the real flavor of the vegetables is retained. If residents do not care then the management of the remaining POT is to be transferred to a central center where it will be managed by the silver generation (the elderly). This can solve the employment problem of the growing silver generation and can complement the characteristics of the system that must be manipulated by hand. The local food is collected by the silver generation to be saved or moved to any place where it is needed. The central center puts their HIVE in the middle of the B-POT project. The role of the HIVE is to repair and maintain the Bee and the POT, as well as to collect the greens. B-POTs are responsible for many roles and will expand your understanding of digital technologies, media, and architecture.

B-POT의 모습을 자세히 살펴보면, 3가지 구성으로 되어있다. 식물을 보관하고 기르는 중심부의 Plant Pocket과 Bee가 POT을 옮길 수 있도록 장착될 수 있는 구멍이 있다. 구멍이 있는 부분을 날개라 하고, 날개는 자유자재의 각도로 변해서 건축물의 다양한 형태를 구성할 수 있게 한다. 또한 POT끼리 서로 지지해주고 받쳐주며 유동적인 건축을 형성한다. Bee가 장착되는 Bee whole은 bee가 잘 빠져나가지 않도록 상층부 부분을 하층부보다 좁게 하였다. B-POT은 집단지성에 의해 움직인다. ICT(Information & Communication Technology)기술과 IOT(Internet of Things)기술의 발전으로 지능을 가진 모듈은 자신이 필요로 하는 부분을 알고 조절할 수 있다. 1개의 B-POT은 2인이 1주일간 먹어야 하는 채소를 제공할 수 있고, 각 가정은 실내의 인공조명을 받은 맛이 없는 식물을 섭취하는 것이 아니라 실제의 햇빛을 받은 싱싱한 채소를 제공받게 된다. 주민들이 관리하지 못하는 경우나 남는 POT의 관리는 중앙센터로 옮겨져 실버세대에 의해 관리되게 된다. 이는 늘어나는 실버세대의 구직문제를 해결할 수 있으며, 손이 닿아야 하는 시스템의 특성을 보완해줄 수 있다. 실버세대에 의해 채집된 로컬 푸드는 저장하거나 필요한 장소로 옮겨지게 된다. 중앙센터는 산하의 작은 HIVE들을 두게 된다. 이 HIVE의 역할은 채소를 채집하는 것뿐만 아니라 Bee와 POT들을 수리하고 관리한다. B-POT들이 수행하는 많은 역할들을 보며 디지털 기술과 미디어, 건축의 이해의 폭을 넓힐 수 있을 것이다.

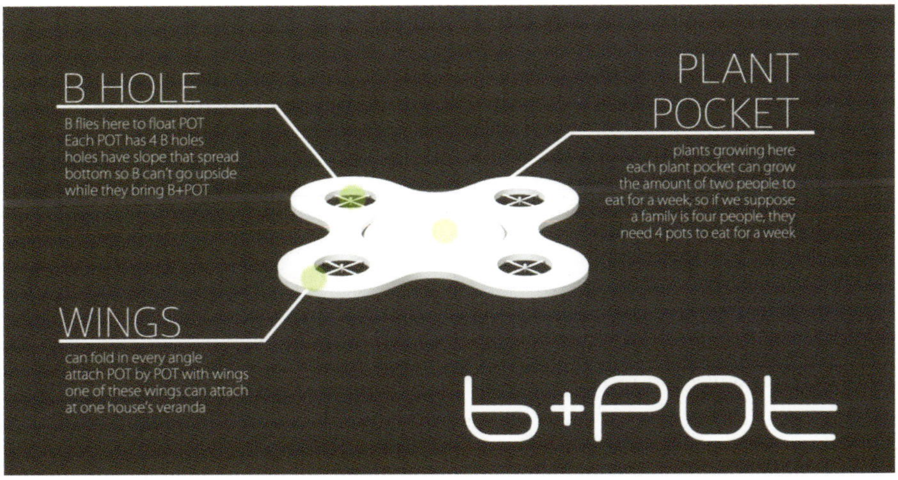

Detail of B-POT

B hole / Wings / Plant pocket

Total view from below
3D printed model

MOVIE MAKING

In the course of making the video, we were trying to express our ideas on the basis of a collective intelligence, and the action of Phototropism and Slime Molds moving in the light of the sun. Because of the particular theoretical background, we can say that we have an advantage, unlike others proposals. This one is based on the movement of the B and POTs, and we tried to make the film with images that would display the incoming future roles needed for a Bio-city. The B moves to find the right light for vegetables. This is to improve the quality of the vegetables and aims to provide the residents of the apartments with fresh produce. The B which drifts in the sky, moves with the necessary solar power. It gives the apartments' residents the wonderful experience of growing their own vegetables and eating food that they themselves have produced. Any remaining amount of vegetable cultivation is sent to the center that controls the B+POT system. Residents will be able to sell their vegetables to the outside. It is proposed as the new idea of city-sufficiency produces added value that the modern city is seeking. In addition, by working through the night and considering the energy that is gained during the day from the media façade, new viewpoints of the city are created, and residents will be able to share within the cultural infrastructure. We have captured forward-looking information with the image of the BEE. This will help you to accurately understand the concept of the B+POT.

영상물 작업을 하는 과정에서 우리는 B가 무리지능과 집단지성을 바탕으로 움직이며, 태양의 빛을 따라 움직이는 굴광성과 점균의 작용을 표현하려고 하였다. 이는 다른 제안들과 달리 우리가 가지고 있는 특별한 이론적 바탕이자 우리의 강점이라 할 수 있기 때문이다. 이를 바탕으로 B와 POT이 움직이게 되며, 움직이며 다가올 미래 Bio-city에 필요한 역할을 영상에 담으려고 했다. B는 하늘을 날며 채소에 적합한 빛을 찾아 움직이게 된다. 이는 채소의 질을 높이고, 아파트의 거주자들에게 신선한 채소를 제공하는데 목적을 두고 있다. 또한 하늘을 날며, 움직이는데 태양열을 통해 움직이는데 필요한 동력을 얻게 된다. 아파트 주민들에게 채소를 자급 자족할 수 있는 기회를 주는 것이다. 채소 재배에서 남는 양은 B+POT 시스템을 통제하는 센터에 보내지며, 주민들은 각자의 채소를 외부에 판매할 수 있게 된다. 이것은 요즘 도시들의 추구하고 있는 부가가치를 생산해 내는 자족 도시의 새로운 제안된다. 또한 야간에는 낮 동안 얻게 된 에너지를 통해 미디어 파사드 작용을 하며, 도시의 새로운 경관을 만들고, 주민들과 문화적 인프라를 공유할 수 있게 된다. 우리는 BEE를 통한 미래 지향적인 내용들을 BEE의 관점에서 영상에 담았다. B+POT에 대한 개념을 더욱 정확히 이해하는데 도움이 될 것이다.

3D PRINT

The 3D printed model of the B+POT project was different from the approaches of other projects. This project was not going to have a solid form of expression because our concept of the B-POT looks like birds and insects that are living in the walls of an existing building. The project initially had an atmosphere of a free and dynamic nature, and modeling this approach of our concept was not easy. The result of the final 3D printed model we decided on was designed to look at the specific architecture of the BEE communities with intelligence, and 3D printing somewhat succeeded in the expression of small units of the B+POTs. What we have been successful in is the overall shape of the cluster form, but it failed to represent the basic form of an architectural scale as the B+POT did not match because the size of the units were too small. Here we realized the limitations of the size that can be represented with 3D printing. Because we had no experience working with a 3D printer, this step was very hard to go through. However, it was a great experience that was uncommon. If we had another opportunity to use a 3D printer, we might be able to obtain a better result by determining the sense of scale for the level and size of the required model.

B-pot 프로젝트의 3D 프린트모델링은 다른 계획들과는 그 접근에서부터 차이가 있었다. 이 프로젝트는 확고한 형태를 가지고 있는 것이 아니며 기존 건축물의 벽에 새와 곤충처럼 서식하는 듯한 모습을 표현해야 하기 때문이다. 애초에 이 프로젝트는 자유롭고 동적인 특징을 가진 터라, 모형 모델링의 접근이 쉽지 않았다. 우리가 결정한 최종적인 3D 프린트 모델링의 결과물은 지능을 가진 BEE의 특정 건축물을 통한 군집 모습을 디자인하였고, 3D 프린트를 통해서 B+POT의 작은 유닛들을 표현하는데 어느 정도 성공하였다. 우리가 성공한 것은 군집 모습의 전체적인 형상이었지만, 실패한 것은 B+POT의 한 유닛의 크기가 너무나 작은 것이기 때문에, 건축적 스케일과 맞지 않아 B+POT의 기본 형태를 표현 하는데 실패하였다. 여기서 우리는 3D 프린트가 가진 표현할 수 있는 크기에 한계를 느끼게 되었다. 3D 프린터를 다루어 본 경험이 없었기 때문에, 더 어려움을 갖는 작업이 아니었나 싶다. 하지만, 그렇기 때문에 신선한 경험이었다고 생각한다. 말로만 듣던 3D 프린트 모델링을 재대로 인식할 수 있는 기회었기 때문이다. 또 다시 3D 프린트를 사용할 기회가 온다면, 필요한 모델링의 수준과 크기에 대한 스케일감 등의 수준을 결정하여 보다 좋은 결과물을 얻을 수 있을 것 같다.

Media wall of apartment

SOCIAL CLOUD

Chloe GIRARD Isabelle KAEKIN LEE Kyungmin NAM Youngju YOON Siu

Social Cloud

Seoul Apgujeong Hyundai apartment / Perspective

In the early 1970s, Apgujeong, where the Hyundai Apartment complex is located, became the most expensive area in Seoul. The facilities are now quite out of date and show the problems of the area. Traffic has been another issue especially the lack of parking spaces. There has also been a failure to merge with the beautiful scenery of the Han River even though it is right at the waterfront. These have been pointed out to be some of the shortcomings of the area. We analyzed the problems and attempted to provide new energy and roles with the Bio digital city concept that we defined.

1970년대 초반, 강남의 노른자위 땅으로 부상한 압구정, 그리고 현대아파트는 예전부터 지금까지 서울에서 부지가 가장 비싼 것으로 유명한 지역이다. 그럼에도 불구하고 20년이 넘는 세월을 이기지 못한 채 낙후된 내.외부 시설은 이 지역의 문제점을 적나라하게 보여준다. 이와 더불어 차량 네트워크가 고려되지 않아 교통이 혼잡하고, 주차시설의 부족으로 인해 난황을 겪고 있는 실상, 바로 앞의 한강이라는 waterfront와 같은 자연 환경을 가지고 있음에도 불구하고, 조화를 이루고 있지 못하는 실상 등의 여러 가지의 문제점들은 현대 아파트가 가지고 있는 단점이다. 우리는 이를 심도있게 분석하고 파악하여 우리가 정의내린 바이오디지털을 가져와 현대아파트 부지에 새로운 활력과 역할을 부여하고자 하였다.

Cloud Control

Bio Digital City / Social Cloud / Cloud / Control Diagram

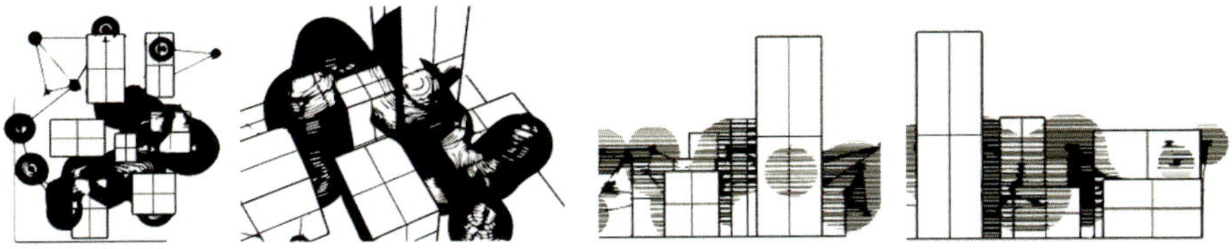

We defined Bio Digital as Bio combined with Digital. Bio and Digital are interpreted separately in their respective fields. The potentials that Apgujeong can have by combining these two concepts and what possibilities the area can pursue were also defined.
"Bio" was perceived as a narrative about 'Vertical Farming' which is the core of urban agriculture, and it could solve the gap between the rich and poor with many possibilities of all-around solutions including the use of software and hardware. 'Digital' was approached as a system to control environments or situations which will be managing the environmental aspects of urban agriculture. We had it play an important role to combine the features from urban areas and nature smoothly.
We specified the discussion described earlier and came to the metaphor of 'Cloud'. Cloud well represents the core of our definition of Bio-Digital. It is flexible and controls the environment by circulating. Such organic traits of 'Cloud' was brought to the site to be actively applied.
We tried to create a circulation - not just a mere environmental circulation but also the circulation in the networks - in the obsolete apartment site with a metaphor of an organic cloud. Going further, it was an attempt to produce circulation and improvement in Seoul by urban networks beyond architecture. We developed 3 opinions from the traits of clouds. First, Cloud by emptying out. Second, Cloud on Vertical farm. And Third, Cloud with bio-control.

우리가 정의한 Bio-Digital은 Bio+Digital이다. Bio와 Digital. 각각의 영역에서 이를 해석하고, 그 속에서 서울 압구정은 어떤 가능성을 지니고 있는지 또한 앞으로 어떤 가능성을 가지고 갈 수 있는지에 대한 물음에서 정의되어졌다. 'Bio'는 도시농업의 주축인 'Vertical Farm'에 대한 이야기로 해석되었고, 이것을 통하여 빈부격차가 심한 압구정의 문제점을 해결하고, 소프트웨어 및 하드웨어를 포함한 전반적인 개선의 가능성을 열어주고자 하였다. 'Digital'은 환경이나 그 상황을 컨트롤할 수 있는 시스템으로 접근하여, 도시와 농업이 가지는 환경적 요인을 컨트롤하고 관리하는 것으로 해석하였다. 이것은 압구정의 도시적 요인과 자연적 요인을 자연스럽게 결합하도록 하는데 중요한 역할을 하도록 하였다.
앞서 말한 이야기에서 의논을 구체화 한 뒤, '구름'(cloud)이라는 Metaphor를 도출하였다. 구름은 우리가 정의한 Bio-Digital의 핵심을 굉장히 잘 함축하고 있는 매개체이다. 구름은 Flexible하며, 환경을 Control하고, 순환시키는 성격을 가지고 있다. 이러한 유기적인 성질이 강한 '구름'을 가져와 이를 부지 내에서 적극적으로 활용하고자 하였다. 우리는 낙후된 현대아파트 부지를 유기적인 구름이라는 메타포를 통하여 순환시키고(단순한 환경적 순환을 넘어서 네트워크적인 순환을 의미) 활성화하고자 하였으며, 더 나아가 건축뿐만 아니라 도시적 네트워크 환경을 구축하여 서울을 순환시키고 개선시키고자 하였다. 오랜 회의 끝에 우리는 구름에서 가져온 특징을 가지고 3가지의 의견을 도출하였다. 첫 번째는 empty를 통한 cloud, 두 번째는 수직농장에 대한 cloud, 세 번째는 bio-control에 따른 cloud이다.

First, Cloud by emptying out

It started with the coexistence of the 'fullness' and 'emptiness' of the site. The Hyundai Apartment complex has the disadvantage of difficult access due to lack of public spaces. This leads to the bigger issue which is lack of access to the Han River and its utilization as an intrinsic advantage. As a solution, we cleared out each apartment up to the 2nd floor to suggest it be used for public space that could be shared with visitors while maintaining the current flow of human traffic. The newly emptied out spaces were provided for residents as a guideline. The current flow of human traffic was guided by various plants for people to adapt to the changes without resistance.

Second, Cloud on Vertical Farm

Apgujeong has a wide entry but this is a disadvantage in the sense of software. So we proposed a 'Triangle parameterize' while maintaining the current shape of the area. There was no biological issues on the site, so we suggested a staircase-shaped terrace while keeping the base triangular. This space is provided to each resident for them to enjoy the family garden and to share with others as well.

첫 번째, 비워짐을 통한 Cloud

이는 사이트에 공존하고 있는 '차있음'과 '비워짐'으로부터 시작되었다. 현대 아파트는 공공부지가 부족하여 외부에서 쉽게 접근할 수 없는 단점을 가지고 있다. 한강이라는 좋은 환경적 요소를 가지고 있음에도 불구하고, 이를 사용할 수 없게 되는 단점으로까지 이어진다. 우리는 이를 해결하기 위해, 일단 아파트의 기존 동선은 유지하되, 각 아파트 마다 2층까지를 비워냄으로써, 외부사람들과 함께 사용할 수 있는 공용 공간을 제시하였다. 비워진 공간은 아파트 주민을 위해 새로운 공간으로 제공되어지는 것을 가이드라인으로 하였다. 그리고 기존 동선에 다양한 조경을 심어 동선을 유도함으로써, 거부감이 없도록 하였다.

두 번째, 수직농장에 대한 Cloud

압구정은 진입동선이 넓어 접근이 용이하지만 쉽게 접근할 수 없는 소프트웨어적인 단점이 존재하고 있다. 그래서 기존의 형태를 유지하는 방향에서 새로운 '트라이앵글 파라메트리라이즈'(parameterize)를 제안한다. 이 장소자체에서의 어떠한 생물학적 문제제시가 없었으므로, 삼각형 형태의 Base를 유지해주고 건물들을 키워 계단식 층계 형태를 제안하여 각각의 테라스가 형성되고 셀들이 각각 거주자들에게 보급되는 형태가 되어서 가족적인 정원을 공용 공간으로 나누는 것을 목적으로 한다.

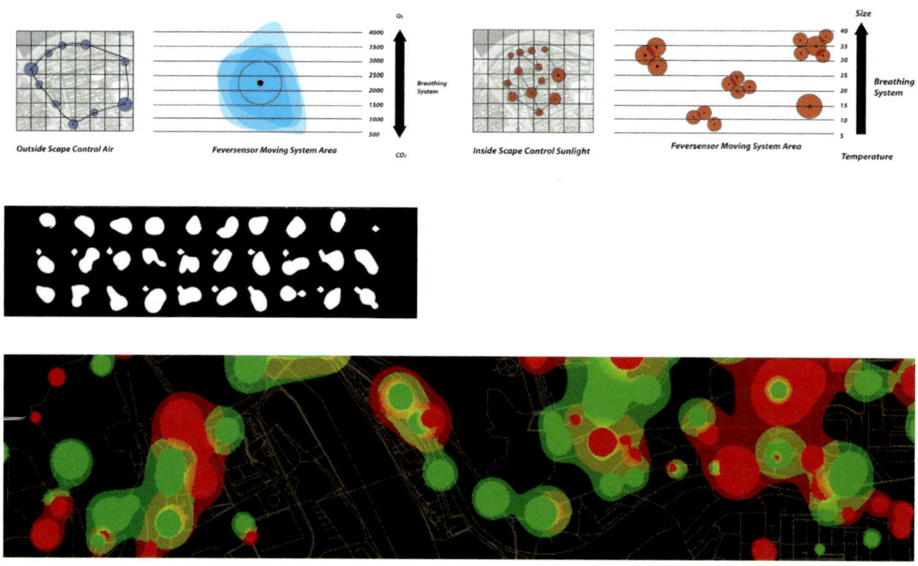

Cloud Diagram

Sunlight control / Air control / Site of Empty cloud / Cloud

Thrid, Cloud with Bio-control

The cloud system described earlier created quite an amount of public spaces by vertical farms at the top of the apartment and large farms at the bottom; and we came to realize that a systemic aspect is required to control all of them. We divided the required aspects into air control and sun-light control and started designing them.

Air-control involves the accumulation of carbon dioxide until it reaches a certain density and then produces oxygen. A media facade was brought into the design to alert citizens.

Sun-light-Control aims to maintain a pleasant internal environment by temperature control. It contracts and expands depending on the temperature while grouping and dividing the spaces. It creates a dynamic space which is also applied to public design areas such as seating and camping areas so that more diversity is brought to the people.

세 번째, Bio-control에 따른 Cloud

앞서 말한 두가지 클라우드 시스템으로 인해 아파트 위의 수직농장과 아래부분의 대형 가든을 통하여 많은 공용공간이 생겼고, 이를 컨트롤할 수 있는 시스템적인 요소가 필요하다고 판단되었다. 따라서 이를 Air Control과 Sun-light Control으로 분류하여 새롭게 디자인하기로 하였다.

우선 Air-control은 이산화탄소를 축적시켜 특정 농도를 넘어서면 산소를 발생시킴으로서 환경을 정화하고, 여기에 미디어 파사드를 도입하여 도시 사람들에게 경각심을 알려줄 수 있도록 디자인하였다.

Sun-light-Control은 사이트의 온도와 관계하여 내부를 쾌적하고 윤활하게 할 수 있도록 함에 목적이 있다. 사이트 내부의 온도에 따라 수축과 팽창을 반복하며, 그루핑하고, 디바인딩되며, 이를 통해 공간이 형성되고 분해되는 다이나믹한 공간을 창출하고자 하였으며, 이것은 공공디자인의 영역에도 포함되어 사람들이 휴식할 수 있는 의자, 캠핑장 등으로 적용되어 풍성한 공간을 연출시킬 수 있게 된다.

Cloud Perspective

Bio digital city / Cloud / Public Design / Perspective

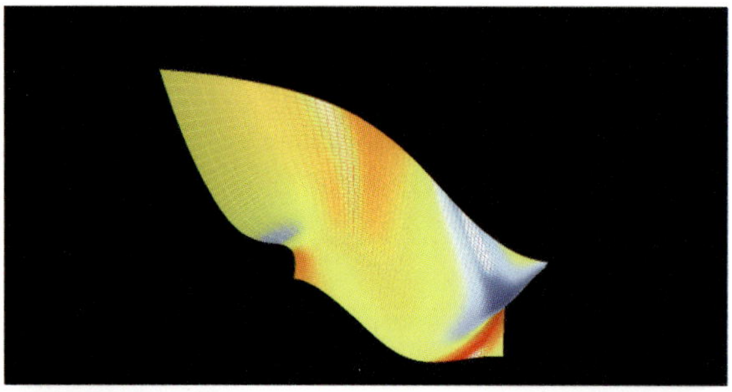

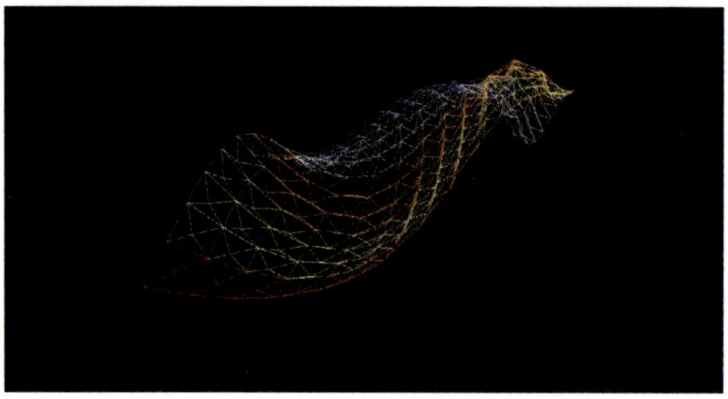

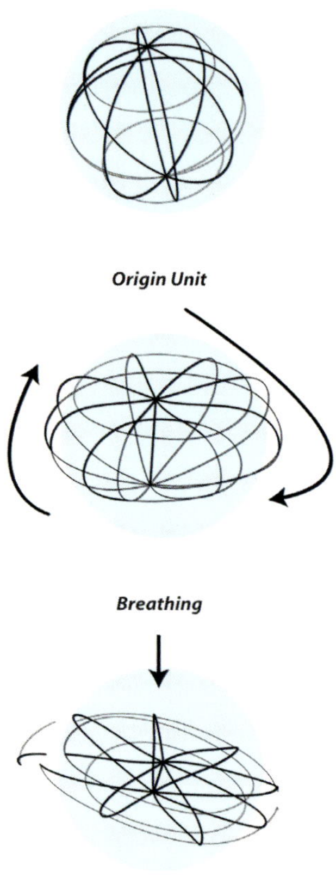

We tried to develop our approach in more detail by using various programs such as Grasshopper, Aftereffect, and Rhino. Putting data into different alternatives and experiments, we shared a lot of in-depth thoughts about how software features can be expressed in a more aesthetic sense. Targeting a video presentation as the final outcome, we decided to produce a documentary film to show the concept and process of our work. We put a lot of effort into the details of the film including creating various stories and recording participant interviews. It was also the team members from far-away France that devotedly cooperated with us and who made this exciting new approaches possible.

작품을 만들기 위해 그래스호퍼, 에프터이펙트, 라이노 등의 다양한 프로그램을 사용하여 보다 더 세심하게 접근하려고 노력하였다. 각 수치를 대입하여 나올 수 있는 여러 가지 대안에 대한 고민을 하여 실험적이면서도 심도있게 접근하였으며, 이와 같은 소프트웨어적인 부분까지도 어떻게 보다 더 심미적으로 표현할 수 있을지에 대하여 많은 고민을 나누었다. 작업의 결과물을 동영상으로 보여주는 것으로 최종 목표를 두고 다양한 의논 후, 우리의 개념과 과정을 잘 보여줄 수 있는 다큐멘터리를 제작하기로 결정하였다. 콘티를 제작하고 수정하여 최종 콘티를 완성하고, 각 참여자 및 조원들의 인터뷰를 녹화하는 등 작품의 디테일을 높이기 위해 많은 노력을 기울였다. 또한 프랑스라는 먼 나라에서 타지로 와서 작업을 위해 기꺼이 노력을 아끼지 않았던 팀원들 덕분에 평상시에 생각할 수 없었던 다양한 접근을 시도할 수 있어서 보다 더 다양하고 재미있는 결과물을 도출할 수 있었다.

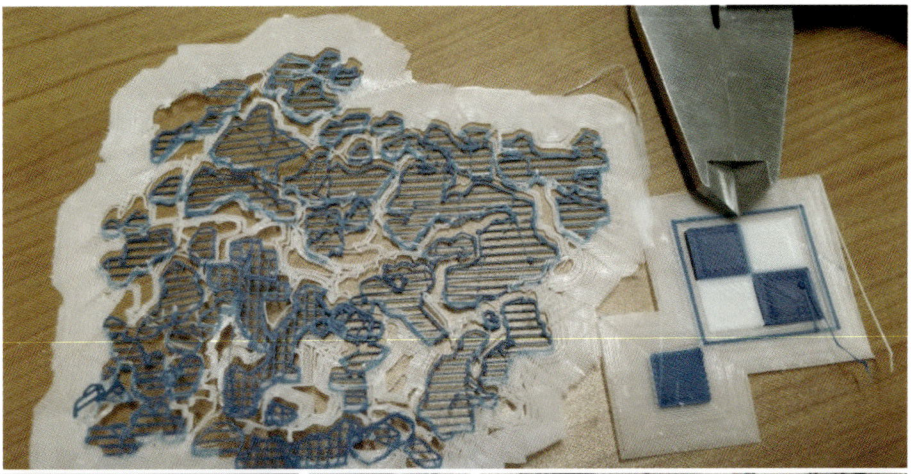

3D Print Model 1

Metaphor / Cloud / Model

3D Print Model 2

Metaphor / Cloud / Base Model

The 3D printing of our work came with a great deal of difficulties. We decided to produce a concept model and an architecture model to be presented together showing how these two operate. We also considered how aesthetic features should be emphasized, led by the projection of a flexible image of cloud. The three-dimensional expression of four sides was significant so we designed each layer to be peeled off. As a result, it turned out to be a perfect model.

3D 프린트 작품을 표현하기에 앞어서 우리의 작품을 드러내기에는 상당한 어려움이 뒤따랐다. 조원들과 의논 후, 컨셉모형과 건축모형을 각각 만들어 이를 함께 보여주고, 두 개의 요소가 어떻게 작용하는지에 대하여 보여주도록 하였다. 입체적으로 표현하는데 있어서 심미적인 요소를 어떻게 부각시킬 것인가에 대한 고려도 하였으며, 이는 구름의 flexible한 이미지를 어떻게 보다 더 잘 형상화할 것인가로 이어졌다. 4면의 입체적 감각이 중요하였기 때문에, 기존틀을 만들어 껍질을 벗겨가는 식으로 제작을 하게 되었다. 결과적으로 완벽한 모형이었다.

T - D

Recently urban agriculture has been recognized as a significant industry to improve the quality of citizens' lives along with bringing attention to healthy food. Frank Llyod Wright suggested 'Broadacre City(1934)' which was an innovative theory and idea that showed the potential of a future agricultural city with cutting-edge transportation technologies. Each family can own an independent farming area and live self-sustainable lives within an architectural environment which suggested true independent living in an urban area. So urban agriculture has been discussed and developed as a significant feature in modern architectural studies.

In team D, we contemplated about the role of urban agriculture in Seoul, the benefits and the potential uses through workshops with students. It started with research questions about urban agriculture and was guided into the following ideas:

1. How to merge digital media technologies with urban agriculture.
2. How to connect physically remote places with diverse functions.
3. How to utilize media as a communication and design method to promote residents' active participation.

We reached the conclusion to select three areas in Seoul for students to propose design ideas appropriate for each region. The efficiency of media and new experiences in architectural spaces in the context of urban agriculture is provided to the students along the way. They were also guided to systematically develop their methodology to establish and realize their ideas.

Amongst the topics of Seoul as a food independent city were: The Edible & Strange City, Seoul and For City Health Seoul. The most useful digital feature was the sharing environment that promoted users' active communication and information exchange. Team A proposed an architectural design with Utopian urban scenarios. A self-sustainable city refers to the system that produces and consumes everything itself. Considering the high population density in Seoul, the research to determine the land size needed to produce the daily amount of food required for a Seoul citizen (based on 2,500 kcal a day standard) was conducted. If the given site (Ichon Apartment complex and the Han River Mansion areas) is not enough, the solution had to be suggested through design ideas. We also spent a good amount of time discussing the sustainable roadmap in an urban context of how residents would be provided with a source and how they would consume the food. Team B proposed an Edible park in the city to design a space to provide new experiences and activities for citizens. Considering the climate and the site's bio-characteristics (the Dangsan factory site and the Lotte car repair sites), the research for the edible garden was performed. It also included the connection amongst open markets, restaurants, and urban farms to consider the economic sustainability as well. The easiest way to be provided with food in the city is the supermarket. But when you look into all the food mileage of most of the products in the market, the price inevitably goes up not to mention all of the emissions made along the way. So mobile farm units, portable space that can be used in each home, were designed and proposed. The site was the Ichon Apartment area where complex residential areas are located.

Team D-1, D-2 and D-3 worked together on the idea of connection through mobiles to provide information about plants, share ideas for farming and distribution - which is a natural form of communication amongst residents through media. Going further from there, the students were guided to consider the media technology center, common and personal farming with architecture optimized for solar energy and energy efficiency. The students' were instructed that the final outcome of the project was to be video presentations with scenarios and storyboards and to deliver a clear message with interesting visual effects. This process itself was a good opportunity for students to learn about presentations of architecture in different formats.

Lastly, urban agriculture has sufficient potential to work as an alternative for the urbanization problems that Seoul is already going through. Especially the landscape of Seoul that is so full of grey buildings due to the rapid population density growth. This landscape issue can possibly be solved by urban agriculture which can provide green views as well as efficient city functions. More than anything, it can contribute to create an eco-friendly Seoul where green plants bring flowers, birds and bees all together.

WOO Junghyun

최근 도시농업은 친환경, 건강한 먹을거리에 관한 관심과 함께 도시민들의 삶의 질 향상을 위해 필요한 중요한 산업으로 인식되고 있다. Frank Lloyd Wright가 제안한 Broadacre City(1934)는 도시 주거공간에 농업공간을 최첨단 교통기술을 이용하여 접목함으로써 미래 농업도시로의 가능성을 보여준 획기적인 이론이자 아이디어였다. 각 가정이 독립된 농경지를 소유할 수 있고 스스로 자급자족할 수 있는 삶을 건축환경과 함께 제공함으로써 진정한 독립된 도심속 삶을 제안하였다. 이렇게 도시농업(urban agriculture)은 현대건축이론에서 중요한 요소로 언급되어 왔고 발전되어 왔다.

D팀에서는 학생들과의 워크샵 작업을 통하여 서울 도심 속에서 도시농업이 가진 역할, 장점, 미래의 활용 가능성에 대하여 고찰해 보았다. 도시농업에 관한 연구질문을 시작으로 첫째, 어떻게 도시농업에 디지털미디어 기술을 접목할 것인가. 둘째, 물리적으로 떨어져 있는 다양한 기능의 공간들을 연결하기 위한 방안은 무엇인가, 셋째, 주민들의 적극적인 농업 활동참여를 이끌어내기 위한 디자인과 커뮤니케이션 도구로서의 미디어를 공간에 활용하는 방안은 무엇인가에 대한 고민들을 학생들이 충분히 할 수 있도록 지도하였다.
그 결과물로 서울내 세 구역을 선정하여 각 사이트의 특성에 적합한 도시디자인계획 및 건축디자인에 관한 다양한 아이디어를 학생들이 제안할 수 있도록 지도하였다. 또한 도시농업을 구성하는데 있어 건축공간에서의 미디어의 효율성과 새로운 경험 제공의 가능성에 대한 아이디어와 구현성을 유추할 수 있도록 방향성을 제시하였다. 그리고 아이디어가 어떻게 적용되고 발전될 수 있는지에 대한 방법론을 설계하는데 있어 체계적으로 발전시킬 수 있도록 지도하였다.

D팀의 각 사이트별로 정해진 소주제: The Food Independent City, Seoul(자급자족 도시, 서울), The Edible & Strange City, Seoul(먹을 수 있는 도시, 새로운 경험을 주는 도시, 서울), For City Health, Seoul(모두가 건강한 도시, 서울)에서 무엇보다 가장 유용한 디지털 환경적 요소는 사용자간의 활발한 소통과 빠른 정보력이 필요한 공유환경이었다. A팀은 자급자족 도시, 서울을 만들기 위해 유토피아적 도시 시나리오를 구성하여 건축적 디자인을 제안하였다. 자급자족 도시란 스스로 공급하고 스스로 소비하는 도시를 말한다. 인구밀도가 높은 서울도시환경을 고려해 보았을 때, 서울시민이 하루(1인당 하루 2,500cal 열량 필요 기준) 동안 소비하는 식량을 생산하기 위해 필요한 토지의 넓이에 대한 조사가 선행되었고, 주어진 사이트(이촌 아파트지구, 한강맨션 아파트 일대)의 용적만으로 충분하지 않을 경우, 그 해결방법을 건축 디자인 아이디어를 통해 제시되어야 하였다. 또한 어떠한 방식으로 소스를 공급받고 생산된 식량을 제공 및 소비할 것인지에 대한 도시적 관점에서의 지속가능한 로드맵을 고민해 보는 시간을 갖도록 하였다. B팀은 도심 속 Edible Park을 제안하여 도시민들에게 새로운 경험 및 체험활동을 할 수 있는 공간을 디자인하였다. 서울의 기후적, 사이트(당산동 공장부지, 롯데자동차정비공장 등)의 생태적 환경을 고려하여, 본 사이트에 다양한 식재활동을 할 때 필요한 도시민이 식용가능한 조경 타입을 조사하는 것이 선행되었다. 또한 오픈마켓 및 레스토랑 등 다양한 서비스 시설들을 도심농장과 연계하여 건축적 디자인을 제안함으로써 경제적으로도 지속가능한 도심농장이 될 수 있도록 다양한 아이디어를 구상해 보도록 하였다. C팀은 서울 시민들이 모두가 건강한 식량을 제공 받기위한 방법론(작은 디자인, 큰 변화)을 구상해 보기로 하였다. 도심에서 가장 쉽고 보편적으로 식품을 제공받는 곳은 슈퍼마켓이다. 하지만 슈퍼마켓에서 제공받는 대부분의 식료품들은 '푸드마일'을 고려해 보았을 때, 운반과정 및 소비과정에서 많은 이산화탄소를 배출하고 결국 운송비용이 실제 식품가격보다 높게 책정되는 경우가 많다. 그래서 각 가정에서 간단하게 설치가능한 디자인 제품이나 이동가능한(Mobile farm unit, Potable space) 디자인을 고려, 제안하도록 하였다. 사이트는 복합주거단지가 위치한 이촌 아파트지구로 지정하였다.

공통적으로 D-1, D-2, D-3 각 팀은 가장 보편적으로 사용할 수 있는 모바일과의 연동을 통해 식물의 정보를 제공받거나 재배 및 유통 방법에 대한 아이디어를 다양한 방법으로 반영하였고, 미디어로 인해 주민들의 자연스러운 소통이 가능하도록 시스템을 구상하였다. 더 나아가 미디어 농업도시 발전을 위해 학생들이 우선적으로 고려해야 할 건축환경으로는 커뮤니티 활동 촉진을 위한 미디어 테크놀로지 센터, 공동 및 개인 농장 및 농업 생태환경 및 에너지 효율성을 위해 적합한 태양고도 및 방향성을 고려한 건축디자인을 제안하도록 지도하였다. 또한, 프로젝트의 최종 결과물인 동영상 프리젠테이션을 위해 각 팀은 중점적으로 전달하고자 하는 메시지를 위한 시나리오를 작성하도록 지도한 다음, 효과적이고 명확한 내용 전달과 흥미로운 영상연출을 위한 스토리보드 작업과 제작을 진행하였다. 이러한 영상콘텐츠제작 프로세스에 대한 이해와 실습은 학생들에게 또 다른 형식의 건축 프리젠테이션 및 아이디어 표현 방법을 배울 수 있는 기회가 되었을 것이다.
마지막으로 도시농업은 서울이 겪고 있는 다양한 도시화 문제 해결을 위한 대안으로서의 역할을 할 수 있는 충분한 가능성을 지니고 있다. 특히, 도시로의 급격한 인구 집중에서 발생한 과밀화의 성장과정에서 발생한 회색 콘크리트 빌딩숲으로 가득한 서울의 랜드스케이프는 도시농업이라는 매체를 통하여 도심의 효율적인 기능 수행과 함께 쾌적하고 아름다운 도심환경 조성이 가능할 것이다. 무엇보다 도시농업을 통해 회색의 삭막한 도시가 아닌 푸른 식물과 꽃, 새, 벌들이 찾아오는 환경친화적인 서울을 만드는 것에 기여할 수 있을 것이다.

AN Na

The study of urban areas has been consistently done. There are various perceptions depending on the researchers. At the beginning of the 2000s, attention was drawn more towards inner development rather than external development. People evolved from survival meals and began to focus more on healthy food which resulted in the growth of related industries. The concept of urban agriculture emerged and drew lots of attention from diverse fields including Architecture.
At the Bio Digital city workshops in Seoul, the members of Team D chose two places out of five to complete 3 projects. Members shared their new perceptions of the city, not to mention they tried hard to present their ideas to the public.
Team D-1 selected Ichon which is a residential area in Yongsan. They set their goal as a self-sustainable city with information media as the project. Team D-2 chose the Dangsan factory site to propose a scenario where farming, commercializing, and consumption are all possible as well as outdoor activities. Team D-3 worked on a project to re-interpret the food mileages on the existing residential areas.
Each team made a proposal using media features to show that media is not just a small part of our lives but it can be beneficial when applied for diverse purposes. It was presented throughout architectural and systematic outcomes of the projects.
The cities in the projects began with a scenario for urban and architectural design and were then guided into actualization along with enormous amounts of sketches, ideas, discussions, and finally, a video.
I believe these ideas have great value as a proposal for future Seoul as we see it, and hope that the research continues through future workshops.

도시에 대한 연구는 오랜 시간에 걸쳐 끊임없이 지속되어 오고 있다. 각기 다른 지점에서 도시를 바라보는 연구자의 관점에 따라 방향과 과정은 매우 다양하다. 2000년대로 접어 들면서 점차 외적인 성장에 대한 관심에서 더불어 내적으로의 성장으로의 관심이 높아졌다. 끼니를 때우는 것을 만족하던 세대를 지나 이제는 건강한 먹거리, 나를 살리는 먹거리에 대한 관심이 높아지고, 그와 주변 산업이 매우 다양하고 빠르게 성장하였다. 이에 도시농업은 개념의 등장과 함께 건축은 물론 여러 분야의 관심을 받았다
Bio-digital city, 서울 워크샵을 통해 D팀의 참여자들은 총 5개의 사이트 중 2곳을 선정하여, 3개의 프로젝트를 진행하였다. 각 사이트에 대한 첫 조사 단계부터 최종 발표까지 참여자들을 그들이 갖고 있는 도시에 대한 새로운 시각을 공유해주었고, 자신들의 아이디어를 대중이 공감할 수 있는 언어로 표현하고자 많은 노력을 쏟아주었다.
D-1팀의 경우, 현재 주거중심지역으로 활용되고 있는 용산 이촌 지역을 사이트로 선정하였다. 그들은 자급자족이 가능한 도시를 목표로 프로젝트에 정보미디어를 적용한 자신들만의 시각을 보여주었다. D-2팀의 경우, 당산동 공장부지를 사이트로 선정하여, 식물의 재배, 생산 그리고 소비가 가능한 형태로 재생산이 되며, 같은 장소에서 아웃도어 활동까지 가능한 공간으로의 시나리오를 제안하였으며, D-3팀의 경우 기존 주거중심 지역에 푸드마일에 대한 재해석을 적용한 프로젝트였다.
각 팀들은 미디어적인 요소를 함께 제안하여, 미디어가 더 이상 생활의 한 분야만이 아닌 전반적으로 도입 가능하며, 보다 사용자들의 생활에 보탬이 될 수 있음을 보여주고자 했다. 그것은 각각의 프로젝트에서 건축적인 디자인에서 시스템적인 요소로까지 적용되었음을 확인할 수 있겠다.
그들의 도시에 대한 시작을 도시디자인과 건축적 디자인을 하나의 시나리오를 구성하고, 구현하는 단계를 경험할 수 있도록 하고자 지도하였으며, 그 결과물은 수많은 스케치와 아이디어 회의 그리고 최종 영상물로 만들어졌다.
우리가 생각하는 미래도시 서울의 모습을 제안함으로써, 그 가치가 충분하다고 생각하며, 또한 워크샵을 통한 연구가 계속되어지길 바란다.

JUNG Hyunyi

Team D's urban-agriculture is a new architectural proposal combining urban space and agricultural function that is connected physically to separated space by integrating digital media technology. A workshop was opened focusing on 'media as a communication tool' to draw active participation from the citizens.
It was oriented to infer ideas and to realize the possibility of how much media can provide effective and new experiences. This was done by forming an urban-agricultural community and searching for a new construction method that proposes an interactive communication technique between space and space, space and citizens, and citizens and citizens through a digital environment. Moreover, it further helped in the systematic development of a methodology on how the ideas can be applied and developed. The most essential digital environmental factors for the sub-themes of Team D were: Food Independent City, Seoul / The City and The Edible, The Strange / For City Health. This sharing environment requires active communication between users and the transmission of fast information. Thus, each team D-1, D-2, D-3 reflected on the ideas of provision, cultivation, and distribution of plant information by interlinking with the most commonly used mobile, and formed a system of citizens that can freely communicate through media.
In addition, the guidance for a video presentation, which is one of the project outputs, was the role of the media tutor. All Team D-1, D-2, D-3 diversely developed the idea as previously mentioned, 'the role of media in urban-agriculture', and each team prepared a scenario for the message they wanted to deliver. Next, they proceeded to work on a storyboard for effective, clear communication of content and interesting visual appearance, and then proceeded with production. For students, the architectural proposal that utilized video was a new approach. The understanding and practical exercise of architectural process and video content production process became an opportunity to acquire new experience.

D팀의 도시농업은 도시공간과 농업기능이 결합된 새로운 건축적 제안으로서 디지털미디어 기술을 접목하여 물리적으로 떨어져 있는 공간을 연결하고, 주민의 적극적인 참여를 이끌어내기 위한 '커뮤니케이션 도구로서의 미디어'에 주목하여 워크샵을 진행하였다.
디지털 환경을 통한 공간과 공간, 공간과 주민 그리고 주민과 주민들간의 상호작용 커뮤니케이션 방법을 제안하는 새로운 건축방식을 모색하여, 도시농업을 구성하는데 있어 미디어가 얼마나 효율적이고 새로운 경험을 제공할 수 있는지에 대한 아이디어와 구현 가능성을 유추할 수 있도록 방향성을 제시하였다. 또한 아이디어가 어떻게 적용되고 발전될 수 있는지에 대한 방법론을 설계하는데 있어 체계적으로 발전시킬 수 있도록 도와주었다. D팀의 소주제인 Food Independent City, Seoul(자급자족 도시, 서울) / The City, The Edible, The Strange, Seoul(먹을 수 있는 도시, 새로운 경험을 주는 도시) / For City Health, Seoul(모두가 건강한 도시)에서 가장 필요한 디지털 환경적 요소는 사용자간의 활발한 소통과 빠른 정보력이 필요한 공유환경이였다. 따라서 D-1, D-2, D-3 각 팀은 가장 보편적으로 사용할 수 있는 모바일과의 연동을 통해 식물의 정보를 제공받거나 재배 및 유통 방법에 대한 아이디어를 다양한 방법으로 반영하였고, 미디어로 인해 주민들의 자연스러운 소통이 가능하도록 시스템을 구상하였다.
또한, 프로젝트의 결과물 중에 하나인 동영상 프리젠테이션을 위한 지도도 미디어튜터의 역할이었다. D-1, D-2, D-3 팀은 모두 앞서 말한 '도시농업건축에서의 미디어의 역할'에 대한 아이디어를 다양하게 발전시켰고, 각 팀은 중점적으로 전달하고자 하는 메시지를 위한 시나리오를 작성하였다. 그 다음, 효과적이고 명확한 내용 전달과 흥미로운 영상연출을 위한 스토리보드 작업과 제작을 진행하였다. 학생들에게 영상을 활용한 건축적 제안은 새로운 접근 방식의 프레젠테이션이었으며, 건축프로세스와는 다른 영상콘텐츠제작 프로세스에 대한 이해와 실습은 새로운 경험을 쌓을 수 있는 기회가 되었다.

WEAVE : WE ASSEMBLED IN THE VENUE

BAEK Jungki HYUN Jinseon JEONG Yoojin KIM Jieun

Joint Farm

Rooftop View / Graphic / 3D Iimage / Common farm / Community

Our design is based on the joint farm program. Each module of space will be weaved together and six weaved modules within a group will be a unit. Like the joint farm system, the module system consists of 6 cubes and includes common space in the center of the cubes. The common space is operated as a core system for communication and transportation.

The residential area is divided into three mass groups to secure a visual axis toward the Han River. These mass groups are lifted up from the ground to procure a large scale farmland and touch down on the ground again towards the Han River. This figure makes it possible to provide a better view toward the Han River in the residential space and helps to utilize sunlight efficiently.

디자인의 기반은 공동농장(joint farm) 프로그램으로, 모든 공간들은 정육면체 형태의 모듈이 모여 서로 엮이며 하나의 큰 유닛을 만들어내는 시스템에 의해 구성되었다. 모듈은 공동농장 프로그램과 마찬가지로, 여섯 개의 유닛이 중심부의 공동 공간을 형성하며 서로 엮여 나가는 형식을 취한다. 중심부의 공간은 커뮤니티의 형성과 이동의 기능을 수행하게 된다.

주거공간은 한강을 향하는 시각축을 확보하기 위하여 세 개의 매스로 나누어진다. 이렇게 나누어진 매스들은 넓은 면적의 농지를 확보하기 위해 지면으로부터 공중으로 들어올려지게 되는데, 한강변을 향할수록 낮아지도록 디자인 되었다. 이러한 형태는 주거공간에서 한강을 향한 더 나은 조망을 확보하는 것을 가능하게 해주며 햇빛의 효율적인 활용을 돕는다.

Joint Farm Module

Joint farm / Unit module / Space Usage / Graphic diagram

Media Application

Media center / Media applications / High-tech / Community center / Cultural life center

Consumption Statas

Crop consumption statas / Farmland scale / Analysis chart / Basic data

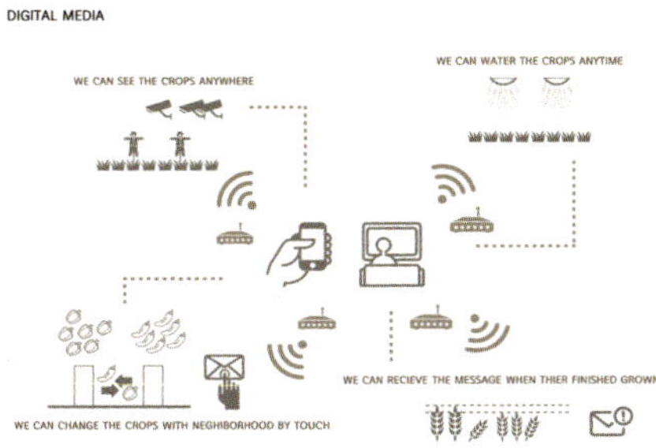

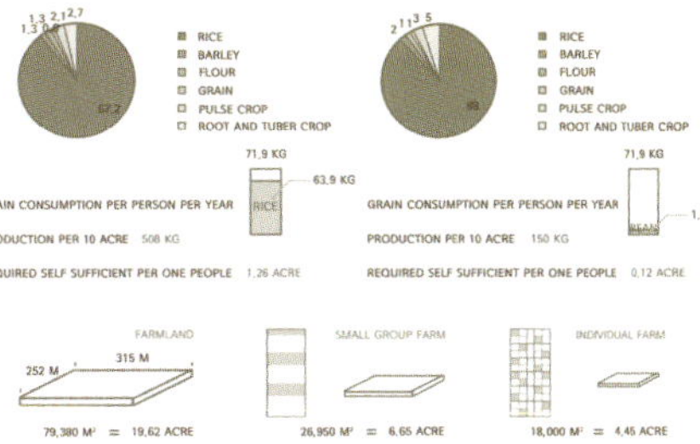

Modular systems are tilted according to the sunlight orientation offering better space characteristices such as water collection and photosynthesis of solar panels on the roof. Solar energy and water aggregation will be used as an energy source to provide a better residential environment.
The common space for a unit consists of six residential spaces facing two common spaces for other units. This, as a result, creates a new community space. People can cultivate crops that can't be grown in an individual or small size joint farm and also can exchange farm products with others.
A complex composed of six residential units includes a common community space for every two units. The ground for the complex is utilized as a large scale joint farm (ground farm) so people can cultivate particular crops which requires a large amount of farmland like rice and wheat. These farm products are stored in the warehouse for the complex and consumed for the dietary needs of the residents.

There are various state-of-the-art media technologies for efficient farm care and community activities in the cube farm complex. People can check the status of their crops or they can take care of their crops in their home by using simple applications so that they don't have to visit the farm directly. The media center is the core facility for the residential complex as it helps to form a community for all the residents. Here, people can gather a lot of information about the status and the distribution of crops in the entire residential area and send messages to each other in order to exchange food.
Moreover, people can receive information regarding their cultural life such as their well-being, lifestyle, various community activities and exhibitions.

정육면체의 모듈은 태양의 방향에 따라 기울어진 형태로 디자인되었는데, 이는 지붕에 설치될 태양열 패널의 에너지 확보와 빗물의 집수에 유리한 공간을 제공한다. 태양열 에너지와 집수된 물은 더 나은 주거환경을 조성하기 위한 자원으로 사용된다.
여섯 개의 주거공간이 엮여 형성된 유닛들의 공동공간은 다른 두 유닛들의 공동공간과 접하면서 새로운 커뮤니티를 형성한다. 형성된 커뮤니티에서 사람들은 개인농장(individual farm)과 소규모의 농장(joint farm)에서는 재배할 수 없었던 식물들을 재배할 수 있으며, 다른 유닛의 사람들과 작물을 교환할 수 있다.
하나의 단지는 총 여섯 개의 유닛이 합쳐져 형성되는데, 결과적으로 그 안에는 두 개의 유닛간 커뮤니티가 존재하게 된다. 이렇게 형성 된 단지의 지면에는 대규모 농장(ground farm)이 형성되어, 넓은 농지를 필요로 하는 벼, 밀 등의 곡식을 재배하는데 사용된다. 재배된 곡물은 단지 내의 창고에 저장되어 단지 내 사람들의 식생활에 사용된다.

큐브 농장에는 효율적인 농장 관리와 주민간 교류를 돕는 다양한 최첨단 미디어 테크놀로지가 적용된다. 이를 통해 사람들은 직접 농장을 방문하지 않아도 간편한 어플리케이션을 사용하여 집에서 작물의 상태를 관찰하거나 관리작업을 수행할 수 있다. 미디어 센터는 주거지구의 중심부에 위치한 핵심적인 시설로, 단지 전체의 주민들의 커뮤니티 형성을 돕는다. 주민들은 미디어 센터에서 작물의 상태와 분포를 살펴볼 수 있으며, 작물교환을 위한 메시지를 손쉽게 주고받을 수 있다. 뿐만 아니라 이곳에서 사람들은 웰빙 라이프 스타일에 관한 정보를 얻을 수 있으며, 각종 전시회나 다양한 문화 활동들을 체험할 수 있다.

As a project for a self-sufficient city, the Seoul Project is an experimental proposal that suggests a utopian way to develop Seoul into an independent city through urban agricultural planning and digital media technology. The site is Ichon-dong, Yongsan-gu, in Seoul and the Han River and various leisure facilities on the south side. Also, the Yongsan Family Park and the National Museum are nearby.

Before starting our project, we studied our site and found out two main issues we had to tackle. First, the lack of public space for residents and second, the buildings that blocked the view of the Han River landscape. Therefore, we realized that we had to consider the visual axis of the site and to make a particular site plan for a large public space. However we couldn't just utilize the potential of the site as a communicative space that provided only an empty place. Also, the existing ideas that are usually expressed as high-rise agricultural plants could block the view of the landscape of the Han River and make it hard to secure public spaces.

So we suggested not just a commercial agricultural plant, but also farmland (Cube Farm) that reveals an emotional space, similar to the one experience in the Korean traditional rural village, to solve these problems and make a self-sufficient city. Through this Cube Farm, we tried to create a public space in a city scale by connecting community spaces including the Han River and other facilities nearby.

The concept for this project is to 'WEAVE' which meant that 'WE Assembled in the VEnue'. We wanted to make people interact together by weaving the spaces that they occupied. Agriculture will also promote their communication as a cooperative common work.

자급자족 도시, 서울 프로젝트는 서울이라는 도시를 도시농업과 디지털 미디어 기술을 통해 자급자족이 가능한 도시로서 발전할 수 있는 방안을 제안하고자 하는 프로젝트이다. 본 사이트는 서울시 용산구 이촌동으로, 남쪽으로는 한강을 접하고 있으며 인근에는 용산 가족공원과 국립 박물관이 위치하고 있다.

자급자족 도시를 디자인하기에 앞서 사이트에 대하여 조사한 결과, 우리가 반드시 해결해야만 하는 두 가지의 문제를 발견할 수 있었다. 첫 번째는 주민들을 위한 공공 공간이 부족하다는 점이며, 두 번째는 한강의 전경이 건물들에 의하여 차단된다는 것이다. 때문에 우리는 시각적인 축을 고려해야만 하며, 넓은 공동 공간을 제공할 수 있는 단지계획이 수립되어야만 한다고 판단했다. 그러나 단순히 비어있는 공간을 제공하기만 해서는 사이트가 가진 커뮤니티 공간으로서의 잠재력을 활용할 수 없었다. 게다가 고층의 공장형 건물을 통해 표현되는 기존의 자급자족 도시 아이디어는 오히려 한강의 전경을 더욱 가로막게 될 가능성이 있었으며 공공 공간을 확보하는데도 어려움이 따랐다.

따라서 우리는 이러한 문제들을 해결하면서 자급자족 도시라는 목표를 달성하기 위하여 단순히 상업적인 식물공장이 아닌, 한국의 전통적인 시골마을에서 겪을 수 있는 감성적 경험들을 체험할 수 있는 큐브형 농장 복합주거단지를 제안하고자 하였다. 또한 큐브형 농장을 통해 주민들을 위한 커뮤니티 공간을 제공하고, 그러한 커뮤니티 공간을 한강을 비롯한 인근 시설들과 연계하여 도시적 차원의 공공 공간을 창출해 내는 것에 중점을 두었다.

이러한, 우리의 생각을 통합시켜준 건축적 컨셉은 WEAVE(엮다)로, 'WE Assembled in the Venue'(우리는 그 장소에서 모인다)라는 의미를 함축하고 있다. 우리는 공간을 서로 엮어줌으로써, 사람들이 서로 소통할 수 있도록 하고자 하였다. 농업은 사람들이 함께 하는 공동작업으로서, 그러한 소통을 더욱 촉진해주는 매개가 되어준다.

Concept Sketch

Concept / Sketch / Drawing / Floating Residence / Tree

Ground Farm

Graphic / Perspective View / Ground view / 3D Image

CUBE FARM PROGRAM

The Cube Farm is managed by various programs that help people communicate with each other. The joint farm system is the basic program of the cube farm. Six people will be in a group and each person will take charge of the farmland each day from Monday to Saturday. On Sunday, they will assemble together to share information and discuss plans for the farm. This helps people share the experience of the joint farm. These small groups can be weaved together and it will make up a large-scale community that connects the city dwellers.

ASSEMLY FARM PROGRAM

The assembly farm program is a support system that helps to build a large-scale community by weaving small sub-groups. The Cube Farm is completed by assembling individual farms, a joint farm and a ground farm. The kinds of crops are differentiated according to the scale and condition of the farm, thereby encouraging people to exchange crops.

큐브팜은 주민들간의 소통을 도와주는 다양한 프로그램들에 의해 운영된다. 공동농장 프로그램은 큐브팜의 가장 기초가 되는 프로그램으로, 여섯 명의 사람들이 한 그룹이 되어 일주일의 기간 동안 공동농장을 함께 관리하게 된다. 일주일 중의 하루는 서로 모여 농장에 관한 정보를 나누고 관리 계획을 세우며 공동농장에서의 경험을 공유하게 된다. 이러한 여섯 명의 소그룹은 자연스럽게 다른 소그룹들과 엮이며 대그룹을 만들어내고, 이러한 대그룹은 주거단지 전체의 소통을 위한 커다란 하나의 네트워크로 발전된다.

조립농장 프로그램은 이러한 소그룹들이 자연스럽게 다른 그룹과 엮임으로써 더 큰 커뮤니티를 형성하도록 돕는 시스템이다. 큐브 농장(cube farm)은 개인농장, 6명의 공동농장(joint farm) 그리고 대규모 지상농장(ground farm)이 모임으로써 완성된다. 이때 심어지는 작물의 종류는 농장의 규모와 조건에 따라 달라지게 되는데, 이는 사람들의 작물교환을 장려하게 된다.

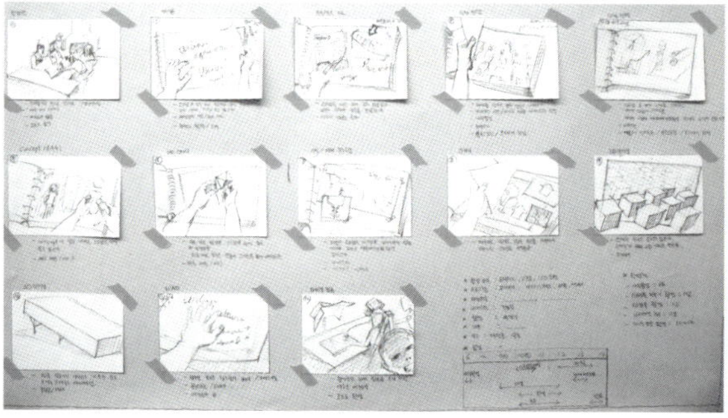

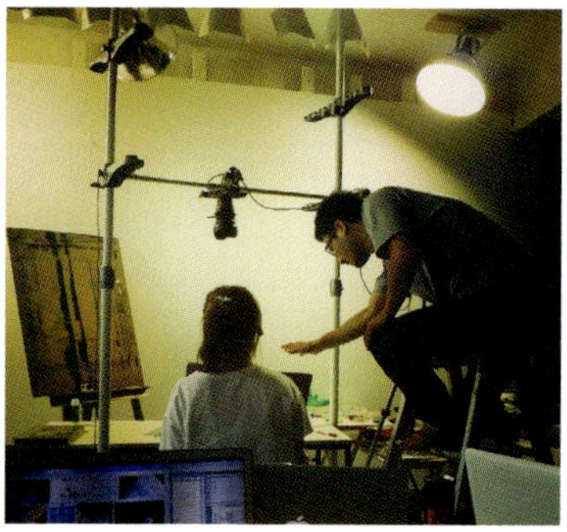

Story Board

Film schedule / Media work / Vedio art / Working plan

Filming

Video art / Directing / Recording / Monitoring

Film work started after deciding on the design concept and the conceptual design. Then we ran with the design development process. Because what we aimed to express was not a high-tech agricultural plant but a complex that can provide a traditional rural experience, the concept for the film worked likewise, and required an analog aesthetic. So, we selected an undecorated shooting method rather than a splendid one.

The method for our shoot, which was expressed in a way similar to turning the of pages of a sketchbook, showed our design process as a recollection of our memories and signified our design concept. Some video editing techniques were applied where things couldn't be expressed by undecorated shooting. It also gave an attractive visual image by providing visual variation and by preventing the film from being monotonous.

영상작업은 디자인 개념이 정해지고 초기 디자인 안이 결정된 이후에 시작되었으며, 디자인 발전 과정과 함께 순차적으로 진행되었다. 우리가 지향하는 디자인이 최첨단 공장형 아파트가 아닌, 향토적인 경험이 가능한 주거단지였기 때문에, 영상 작업 개념 또한 그와 마찬가지로 아날로그적 감성을 가져야만 한다고 생각했다. 그래서 우리는 화려한 영상 연출 대신, 특별한 꾸밈없는 영상 촬영 방법을 택했다.

스케치북에 아이디어를 표현하고, 그것을 한 장씩 넘기며 기억을 추억하듯 디자인 프로세스를 자연스럽게 보여주는 방법은 우리의 디자인 개념을 잘 나타내 주었다. 단순한 촬영작업만으로는 표현하기 어려운 부분들에는 약간의 편집 기술들이 적용되었으며, 이는 영상이 단조로워지는 것을 방지하면서 시각적인 변화를 통해 흥미를 줄 수 있도록 하였다.

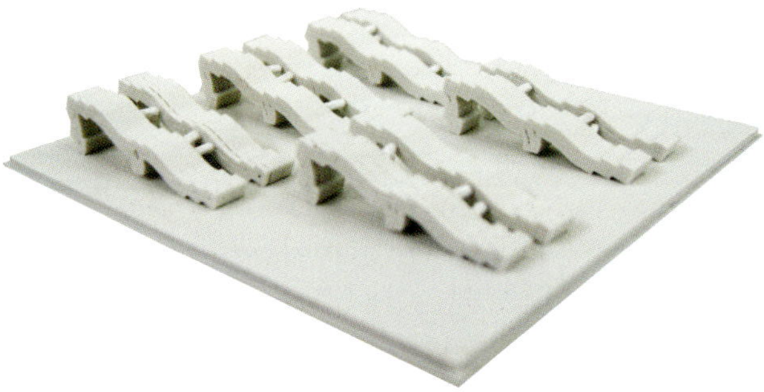

3D Printing Model

Physical model / Mass model / 3D printing / Bird eye view

Weave Complex

Cube farm / Linear farm / Bird eye view / 3D image / Graphic

The 3D printing technology which I experienced for the first time via this project, possesses much more possibility than I ever imagined. The main principle of the 3D printing operation is to lay a thin plastic line upon another so that we can create a very detailed model by controlling the thickness of the plastic lines. If we had made small scale models by hand as is the traditional way, we might have had many problems. However by using 3D printing technology, we could make the model in a very detailed way and in addition, economize the entire working time so we could work on other things while waiting for the 3D printing. Our concerns regarding some curves in our design were also solved as the 3D printing worked seamlessly. Overall, it gave us a great deal of satisfaction.
Another powerful merit of 3D printing technology was its compatibility with other architectural 3D programs. We could make progress in our 3D modeling for rendered shots in a collinear process for the 3D printing, without having to make any effort to do the same kind of work with other programs.
Of course there are still some limitations in 3D printing, such as the problems that come from angles or the points of support for the model. However, it is clear that it possesses much possibility to make the architectural design process richer in terms of workability. By embodying typical architectural models and other works from abundant imagination, 3D printing technology will make architectural content much more luxuriant.

이번 프로젝트를 통해 접한 3D 프린팅 기술은 상상했던 것보다 더 많은 가능성을 가지고 있는 것 같았다. 3D 프린팅은 얇은 플라스틱 선을 겹겹이 쌓아 매스를 만들어가는 원리에 의해서 작동하는데, 쌓여지는 선의 굵기를 조정하면 더 세밀한 형태까지도 만들어낼 수 있다. 때문에 작은 스케일의 모형의 경우, 기존의 방식대로 손으로 작업했다면 어려움이 많았겠지만, 3D 프린팅 기술을 활용함으로써 세밀한 부분까지 정교하게 작업할 수 있었으며, 프린팅이 진행되는 동안 다른 작업을 수행할 수 있어 전체적인 작업시간까지 절약할 수 있었다. 또한, 우리가 작업했던 디자인에 곡선이 다수 포함되어 있었기 때문에, 모형 제작에 대한 고민이 많았었으나, 곡선부분 또한 매끄럽게 제작되어 상당히 만족스러웠다.
라이노 등의 일반적인 3D 프로그램과도 호환이 된다는 점 또한 3D 프린팅이 가지는 강력한 장점인 것 같다. 다른 프로그램을 통해 또 다시 같은 작업을 반복할 필요 없이, 3D 랜더링을 위한 모델링 작업을 진행하면서 프린팅 작업을 시작할 수 있기 때문이다.
모형의 각도나 지지점에 따라 프린팅이 어려운 부분도 있다는 한계점이 존재하기는 하지만, 작업의 능률을 고려해 보았을 때, 건축디자인의 과정을 더욱 수월하게 해 줄 가능성이 풍부하다는 것은 분명하다. 수작업으로는 제작하기 어려웠던 비정형의 건축 모형들의 제작은 물론 더 다양하고 풍부한 상상의 산물들을 현실로 구현해줌으로써, 3D 프린팅 기술이 건축을 더욱 풍부하게 만들어 줄 것이라는 생각이 든다.

SEED TERMINAL

KIM Dongyun KIM Jayeong NA Karam PARK Harang

Seed Terminal

Dangsan, Seoul / Rooftop
Perspective

In contemporary times, with the growth of environmental issues, we distrust the distribution process, GMO plants, and fear the ingredients of the food we eat. However, consumption of vegetables maintains a high level of importance as indispensable elements of a citizen's meal. Accordingly, the number of urban farmers who want to grow their own plants is increasing. At first, it seemed like just a trend but now the idea of urban farming is growing in to a kind of industry which is necessary to improve health, our environment, and quality of life for our citizens, such as education and enhancing the recovery of our communities. Accordingly, perhaps out of necessity, with architectural proposal and digital media, the 'Self-sufficient city', Seoul project suggests an alternative future city which functions self-sufficiently, not demandingly. We propose an urban 'edible park' which provides new experiences and healthy food.

The distribution center is located in the Dangsan industrial site. Next to this site, is Rodle Avenue and the Olympic Expressway which is a major road in Seoul. So we can say that this site has very good traffic conditions that radiate from the site into Seoul. Also, the Han River flows near this site which can be seen as an advantage if it becomes a park. With the remarkable condition and originality of this site, we thought we would be able to design an economical and sustainable park that included not only simple things like open markets and restaurants. However, we cannot expect to develope an urban farm that has strength of access and distribution because this is just a changing location of a distribution center. But depending on "what we distribute", the result can change. Consequently, we came up with a sustainable idea - the consumer produces something out of what they get from this park - and that would be 'the seed' which is the basic source for growing vegetables. Therefore, a real 'edible park' on this site acts as a 'SEED TERMINAL' which is an important strategic point of urban farming in Seoul.

Prior to this proposal, the most important work was analysis of the site and the study of precedent cases, high demanding vegetables, comestible plants and functions. Because we cannot grow every type of vegetable on an urban farm, study of vegetation which can be easily grown in a rough environment was inevitable. Moreover, taking advantage of not only the location but also the position or role of the distribution center, we discussed how to support urban farming in Seoul. Basically under the precondition of keeping the role of distribution, the scheme was utilizing a large space with cultivation and packing of seeds to be distributed in the process of urban farming.

Selecting potential points of green public spaces in all 25 districts of Seoul was also an important study point. Not just choosing green space, it also has to satisfy the criteria such as traffic, passenger's access, adequate area and so on. As connection among edible parks shows real development possibility, information about the district such as area, population, and demands was highly essential. Since we suggest the key points of urban farming, relation is an important element. So, we analyzed data which became the basic design scheme. The final draft was formed from a correlation of this data.

환경적인 문제와 더불어 유통과정에 대한 불신, 유전자 변이 식물 등 현대인들이 공급받는 식재료에 대한 불안감은 어제, 오늘의 일이 아니다. 하지만 채소는 도시민의 식탁에서 빠질 수 없는 요소로 소비량은 늘 높은 수준을 유지하고 있고, 그에 따라 도시의 텃밭이나 주말농장에서 내 손으로 신선한 채소를 기르겠다는 도시농부가 늘게 되었다. 도심농장이 처음에는 마치 한 때의 유행인 것처럼 여겨졌으나, 이젠 먹을거리는 물론 건강과 환경개선 및 교육과 공동체 회복 등 도시민들의 삶의 질 향상을 위해 꼭 필요한 산업으로 바뀌고 있다. 자급자족 도시, 서울 프로젝트는 이러한 사회적 필요성에 따라 건축적인 제안과 디지털 미디어를 통해, 더이상 공급받기만 하는 도시가 아니라 자급자족이 가능한 도시로서 발전할 수 있는 방안을 제시하고자 한다. 우리는 도심속 Edible Park를 제안하여 도시민들에게 새로운 경험, 체험활동과 동시에 건강한 식량을 제공할 수 있는 공간을 디자인하고자 한다.

본 사이트인 당산 공장부지는 현재 물류센터가 중앙에 자리잡은 곳으로, 바로 앞에는 노들로와 올림픽대로 등 주도로가 있어 서울의 각지로 뻗어나가기에 좋은 교통의 요충지라고 할 수 있다. 당산 공장부지가 본래 사용되고 있던 용도와 앞서 말한 사이트의 조건들은 '유통'에 최적화된 것으로 그 외에도 바로 앞에 마주하고 있는 한강 등 공원으로서 장점이 될 수 있는 조건을 갖추고 있다. 유통이라는 특징이 두드러지는 대지의 조건을 통해, 단순히 Edible Park를 오픈마켓이나 레스토랑을 통한 식재료 순환과 서비스 시설이 제공되는 곳으로서만 바라보는 것이 아닌, 경제적으로도 지속 가능한 도심 농장을 가능하게 할 수 있다고 생각하였다. 하지만 단순히 접근하기 좋다는 유통의 장점만으로는 도시농업의 발전을 기대하기 힘들 것이다. 그것은 도시가 늘 공급받는 위치에 있던 과거와 달라지는 것이 없어지게 되기 때문이었다. 하지만 무엇을 전달하느냐에 따라 결과는 달라질 수 있을 것이다. 이 곳에서 전달되는 무언가를 통해 수용자가 다시 생산할 수 있도록 하는 것- 여기서 우리는 도심농장에 필요한 가장 기본적인 '씨앗'을 떠올렸다. 씨앗은 모든 채소가 나기 위한 원천이 되는 것, 즉 이 부지의 Edible Park는 서울 도시농업 발전의 요충지, 'SEED TERMINAL' 로서 역할할 수 있게 하고자 하는 것이다.

아이디어 제안에 앞서 우선적으로 중요하게 생각했던 것은 도시농업의 선례와 더불어 도시민들의 수요가 높은 채소, 식용가능한 조경타입과 Edible Park가 갖춰야 하는 기능, 그리고 사이트에 대한 철저한 분석이었다. 모든 채소가 씨앗을 얻기에 용이한 것은 아니었기에 도심이라는 조건에서 사람들이 많이 찾으면서도 쉽게 씨앗을 얻을 수 있는 채소는 무엇인지에 대한 정보는 필수적인 것이었다. 또한 사이트를 단순히 위치적인 조건으로서만 이용하는 것뿐만 아니라, 사이트가 가진 기존 물류센터로서의 역할도 함께 가져가 그것이 어떤 점에서 도심농장에 도움을 줄 수 있을지 생각해보았다. 기본적으로는, 계획상 물류센터를 무시하지 않고 유지한다는 조건 하에, 그 건물의 대공간을 공원 내에서 씨앗을 유통하기 위한 재배시설과 포장시설로서 활용하고자 하였다.

서울 25개의 구마다 잠재적 공공 녹지 공간을 선정하는 것 또한 중요하게 선행된 작업이었다. 단순히 녹지만을 선정한 것이 아니라, 그 안에서도 사람들이 접근하기 좋은 곳이면서도 적당한 크기를 갖춘 곳 혹은 녹지가 아니어도 당산부지와 비슷한 조건을 가진 곳도 잠재적 거점이 될 수 있었다. 선정된 공간과 SEED TERMINAL 간의 관계가 도시농업의 발전 가능성으로 이어지는 만큼, 거점의 해당 구 면적과 거주민의 수 등, 수요량과 연관된 정보는 중요한 요소였다. 그리고 이 정보들을 알고 지나가는 과정의 일환으로서 끝내는 것이 아니라, 계속해서 끌고 나가 이 정보들을 어떤 질서로, 무엇에 무게를 두면서 연결해야 하는지 고민하였다. 도시농업 발전의 요충지를 계획하는 만큼, 서로간의 관계성은 중요한 것이었고 그것을 디자인의 모태로 삼았다. 최종안은 결국 그 정보들이 서로 연관성을 지니면서 연결되었을 때 도출된 결과물인 것이다.

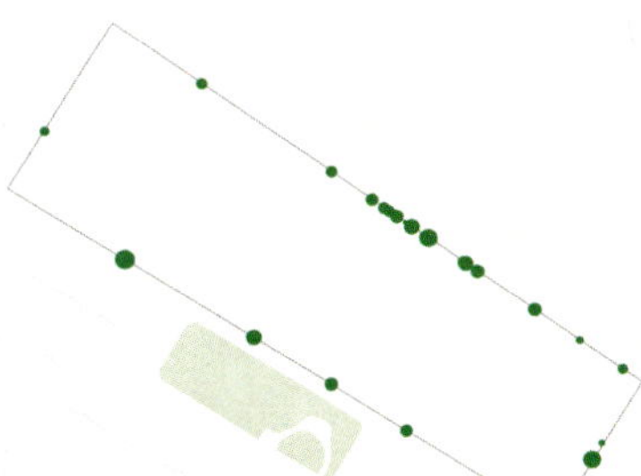

Concept Process

Network system / Diagram

The SEED TERMINAL which is the base point of urban agriculture, starts with the formation of an imaginary network line. This line is based on the connection between the base point in Dangsan and the key points of the districts in Seoul. When this network extension stretches to the key point, those connection lines and site boundary intersect. These intersection points represent each district. According to the required conditions of the district's population density and the space needed, the possibility of an agricultural theme park will be different, and the intersection point's intensity is determined by that information. Therefore, intersections are the key points where possibilities represent potential of urban agricultural growth.

Changing direction, the points form a relation system inside the site. Including their information, the points are connecting. This relationship between the points is described by trace. If the intensity of a point is strong, trace of the network will be wide and thick. In other words, the possibility of urban agriculture in Seoul is condensed into this network. If intensity of the network becomes stronger, the park needs more space. So concentration of trace is an index which indicates the demand for space. To express those connections more efficiently, when melting the trace into the site, the trace will maximize and transform into shape. Finally, the condensed network shapes the landscape of this site. This landscape is visualized by the relation to potential spaces in Seoul. To make use of the formed landscape shell, inside space will be made available to a service facility and a vegetable cultivation facility. We also considered substantive structure issues throughout the ribbed wood frame and steel pipes.

도시농업 발전의 기점이 될 SEED TERMINAL은 당산 부지를 원점으로 삼아, 앞서 선정되었던 서울의 각 잠재적 거점과 가상의 네트워크 선을 형성함으로써 시작된다. 각 거점으로 뻗어나간 네트워크의 연장선은 사이트의 경계와 만나면서 교차하게 되는데, 이 때 남게 된 교차점들은 각 거점의 대표점이 된다. 이 대표점들은 해당 구의 인구 밀도와 공간의 필요 정도에 따라 그곳에 생기는 도심 공원이 활성화 될 가능성이 높은 만큼 내적 강도를 달리 하였고, 이를 그곳에서의 서울 도시 농업 발전의 가능성을 나타내어주는 지표로서 활용하였다.

사이트 외부로 향하던 네트워크를 이제는 방향을 달리하여 내부로 집중시켜 대표점간의 관계를 형성해보았다. 이 때 조건에 따라 강도를 달리 했던 대표점은 네트워크를 형성할 때에 그 성격을 유지하며 연결된다. 이 관계는 점 간의 네트워크가 남기는 흔적으로 표현이 되는데, 강도가 높은 대표점이 연결될수록 넓고 진한 흔적으로 남겨 활성화 될 가능성이 높은 Park 사이의 관계가 더욱 강해질 수 있다는 정보를 시각화해주게 된다. 즉, 이 흔적은 각 지역의 Edible Park가 서로 연계되는 가능성을 응축한다. 네트워크의 강도가 강할수록 많은 공간을 필요로 하게 되는 만큼 흔적의 진하기는 곧 공간의 필요정도를 나타내는 지표가 된다. 그 관계를 효과적으로 표현하기 위해 흔적을 부드럽게 주변에 녹여냈을 때, 그 형태가 극대화된 모습은 사이트 안에 지형을 형성하게 된다. 이는 당산부지와 서울 각 거점 사이의 네트워크 가능성을 시각화한 형태가 되는 것이다. 형성된 지형을 활용한 쉘로 내부의 공간을 서비스 제공 시설과 채소 재배 시설로 활용할 수 있을 것이다. 쉘의 형태를 따라 일정한 간격으로 배치된 프레임과 이를 연결해주는 파이프를 통해, 실직적인 구축 또한 고려하였다.

Rooftop View

Seed Terminal / Farm grass on rooftop

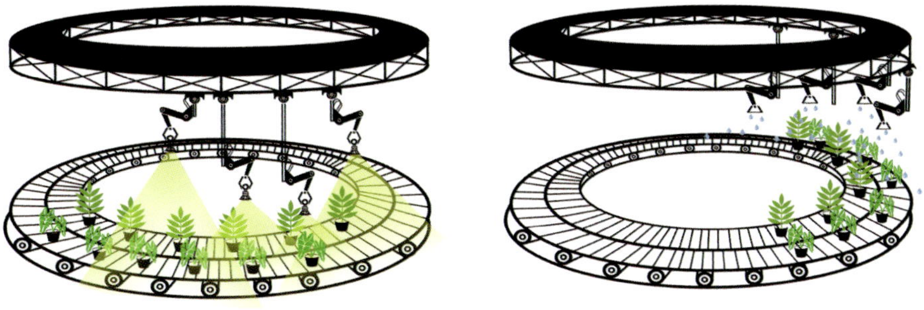

Auto Cultivation

Digital System with Auto Cultivation / Diagram

Considering the shape of the landscape, the rooftop has various gradients and the inner space has an irregular area. So, we planned our program to consider those factors. At the highest slope on the roof, 'Frame Glass' will be used so that sunlight comes into this building. The cultivation room and laboratory will be placed where the glass size is the widest. Rest spaces will be arranged within an open market or restaurant so that people can feel the natural light. In areas where the slope gradient is about 35~65%, we will plan to place a 'Green Landscape'. Through this grass zone, it will function as a park. Where the slope gradient is lower, (about 15~35%), cultivation zones will be placed so people can harvest vegetables there. Those vegetables are supplied straight to the open market and restaurants so people can eat fresh food. The rest of the districts have the lowest level, therefore we plan to make a trail so people can walk on the roof and when they are coming up this road, they can see the Han River in front of the SEED TERMINAL. Finally, following the trail, a café and resting areas are placed.

These days, everyone commonly uses a smart phone; therefore the SEED TERMINAL introduces a new digital system for using the Edible Park more easily. Through an application, people can get information about vegetables, check the process of cultivation in real time, and also order seeds very easily. In each season, this system notifies the people as to which types of vegetables are available. Utilizing the original usage of this site, the SEED TERMINAL uses a conveyer belt system. This system is applied by an algorithm of factory. When vegetables move along the conveyer belt, they are checked to make sure that they are in good condition, so vegetables can be cultivated with few staff. This is a suitable system for the SEED TERMINAL to use its functionality as the base point of urban agriculture in Seoul, which has to ensure supply in a short time.

지형을 형성하게 되는 만큼, 지붕의 경사도는 다양해지고, 내부 공간의 크기가 균일하지 않은 것을 고려하여 프로그램을 계획하였다. 가장 높은 경사도를 갖는 구역에는 지붕에 FRAME GLASS를 설치하여 내부에서는 천창으로부터 떨어지는 빛을 경험할 수 있게 되는데, 그 크기에 따라, 많은 빛이 들어오는 곳에는 식물이 재배되는 실과 연구실을 배치하고, 작은 곳에는 공간을 이용하는 사람들에게 천창으로부터 떨어지는 빛을 경험할 수 있도록 RESTAURANT 등 일반인이 사용할 수 있는 프로그램을 배치하였다. 35~65%의 경사를 갖는 곳에는 녹지 조경을 통해, Park로서 기능할 수 있는 조건을 갖추었고, 15~35% 경사가 있는 곳은 도심농장에서 지을 수 있는 식물을 재배하여 Park 내의 오픈마켓 옷은 레스토랑으로 바로 공급될 수 있도록 하였다. 가장 낮은 경사도를 갖는 곳은 사람이 진입할 수 있는 높이와 함께 고려하여, 멀리 한강을 내다볼 수 있는 조망이 가능한 산책로와 함께, 내부로 진입하였을 때 바로 이용할 수 있는 cafe, 쉼터와 같은 프로그램을 배치하였다.

모두가 흔히 사용하는 스마트폰 어플을 통해, Edible Park를 이용할 수 있도록 계획하였다. 어플을 통해 원하는 채소에 대한 정보를 얻고 주문하게 되면, 그를 재배할 수 있는 씨앗을 제공해주거나 혹은 Edible Park 내에서도 생산되는 정도에 따라 일정량은 개별 주문을 통해 직접 재배한 채소의 형태로 제공해줄 수 있는 시스템이다. 어플을 통해 채소가 어떻게 재배되는지 실시간으로 확인할 수 있고, 계절마다 제철인 채소가 무엇인지 주말농장을 이용하여 씨앗을 재배할 때 관련된 정보 등을 얻을 수 있다.
기존 사이트가 공장부지였던 것을 활용하여 컨베이어 벨트를 이용하였다. 공장의 알고리즘을 이용한 시스템으로 컨베이어 벨트를 따라 이동하는 채소의 상태를 체크한 뒤에 갖춰야 하는 과정을 순서대로 배치하여 많은 인력을 들이지 않고도 상당량의 채소를 재배할 수 있도록 하였다. 이는, 도시농업의 거점으로 역할하기 위해 많은 양을 구비해놔야 하는 SEED TERMINAL로서 충분히 역할하기 위해 적절한 시스템이 될 것이다.

Storyboard

Concept image / Illustrator / Process Slides

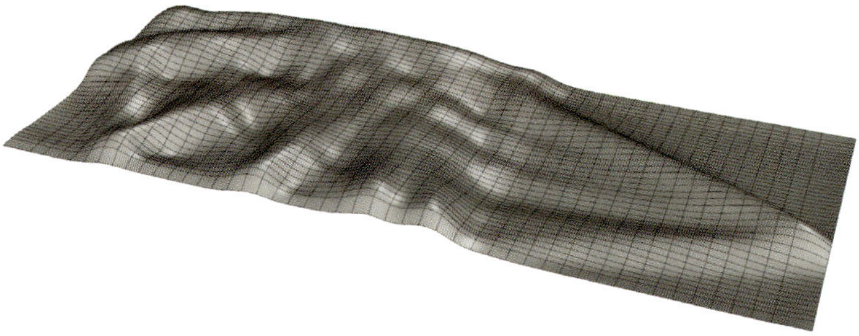

Concept Model

3D modelling / Rhino / Concept model

The whole project developed differently with the general process such as developing our concept model or proposing various alternatives. We would rather not make a form subjectively, and instead, deduce results through the analysis of geographic data and statistics. The shape is related to this data. Showing the process naturally was a problem for us. To solve this, connected and intersected points of information was important. We concluded that our work and study were almost like a text type, so we had to describe the diagram based on texts. In terms of video producing, we wanted to express rather than just create a list of final images, the flow of our process along with a story line. Then, we thought of the image of a dandelion because it spreads its seeds following the wind and then, the seeds take root in the ground. After rooting, seeds will grow into other dandelions and the process starts again. So we compared this spreading process with our concept of the SEED TERMINAL. In 3D modelling, we used Rhino and Grasshopper so that we could transform the shape and study the structure of the system.

전반적인 작업은 초기의 개념 모델이 여러 대안으로 다시 제시되고 발전되는 형식과는 다르게 진행되었다. 형태를 주관적으로 만들어내기보다 지리적인 혹은 통계와 같은 정보들이 서로 지닌 연관성을 이용하여 도출한 것이었기 때문에, 그 과정을 자연스러운 하나의 흐름으로 보여주는 것이 보다 큰 과제가 되었다. 이를 위해서는 활용했던 정보들이 연결되는 접점이 중요한 것이었는데, 이에 관해 논의되어던 것은 텍스트의 성격이 강한 것이어서 이것을 다이어그램으로 보여주기 위한 작업이 중요하게 진행되었다. 영상제작에 있어, 그동안의 과정을 이미지화한 것들을 단편적으로 나열하기보다 그 흐름을 이해해줄 수 있는 스토리 라인을 따라 표현하기 위해, 당산 부지를 기점으로 서울 각 거점에 도시농업을 퍼지게 한다는 SEED TERMINAL의 개념을 씨앗에서 자란 후에 자신도 씨앗을 품은 꽃이 되어 바람을 타고 이곳 저곳에 뿌리를 내리는 민들레의 모습에 비유하였다. 라이노와 그래스호퍼를 이용한 모델링은 최종적으로 형성된 지형을 쉘의 형태로 변형한다거나 구조에 대한 연구를 하는데 이용하였다.

Building Section

Seed Terminal /
Section Diagram /
Program Diagram

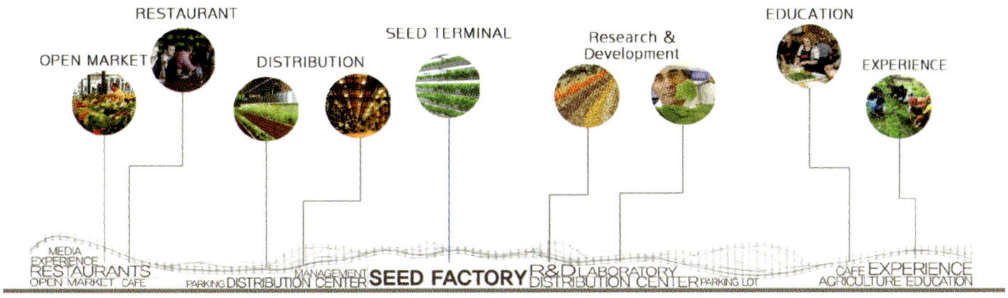

Because each printer has a different way of printing data, even if we had some experience using a 3D printer, it was not that easy. In this workshop, the printer had a FDM method which uses threads of plastic. When the printer put out this thread, it melts the plastic at the edge and builds up following the shape. This method was different for us because we experienced some problems completing our building shape. When we made the model, our first plan was to print it in 3 parts - frames, structures and path. But these were too big to print with the FDM printer - their probable size is 210x150x100mm. So we had to divide our model in half. The FDM printer can't print anything under 0.3mm thickness; therefore we also had to regulate the scale of our model's frame and check it again and again. As previously stated, the output system of the FDM printer has some limitations. For example; when a curved surface of a model is over 45 degrees, the model needs some additions to support its structure. These supports can make the model look messy if we were unsuccessful in our attempts to control the outcome. So we decided not to focus on little details like the frame. Our plan changed into making a solid mass, instead of an overall curved surface model. If there were some parts that fell apart in the 3D modeling process, then the resultant model would collapse. So, before handing over the file, we checked everything very carefully. Most of all, we had to print within a deadline. For that reason, we controlled the density of our model in order to reduce the printing time. Nevertheless, if we made this model ourselves, by hand, we would be unable to express the curved surface, which is the most important design element of our building. Thus, the result was very satisfying and was a pleasant accomplishment for us.

3D 프린터를 사용해본 경험이 몇번 있었기 때문에 생소하다거나 낯설지는 않았지만, 프린터마다 출력되는 방식이 달랐기 때문에 쉽게 진행된 것도 아니었다. 이번 워크샵에서 사용하였던 프린터는 FDM 방식으로 플라스틱 계열의 실을 내보내고 끝부분에서 녹여 적층되며 올라가는 것이었는데, 이 방식은 우리가 만들어낸 형태를 완전히 출력하기에는 조금 어려움이 따르는 것이었다. 모형을 만들 때 처음의 계획은 곡면을 세로로 나눈 프레임들과 가로로 프레임들을 엮어줄 구조체들, 그리고 지붕 상부의 path 를 각각 따로 출력하여 합치는 것이었는데, 제공되는 3D 프린터의 출력 가능한 크기(210x150x100 mm)가 모형을 한번에 출력하기에는 작아 2개 부분으로 분리해야 하는 수 밖에 없었다. 그 밖에도, 두께 0.3mm 이하로는 출력이 불가능했기 때문에, 초기 모델링의 스케일을 조절하면서 출력이 가능한 크기를 재차 확인해봐야 했다. 또한 앞서 말했던 프린터의 적층 출력방식은 곡면의 각도가 45도 이하가 되면 구조적으로 쌓일 수 없게 된다는 한계가 생겨 프린터 내에서 자체적으로 출력되고 있는 매스를 지탱해줄 서포터가 생성되었는데, 이 서포트들은 후처리가 미흡할 시 형태를 매끄럽게 나타내주질 못하고 조금 지저분하게 만들 수도 있었기 때문에, 작은 디테일 보다는 전체적인 형태를 보여주는 방향으로 계획을 변경하여 쉘보다는 솔리드한 매스를 보여주는 방향으로 계획을 변경하였다. 프린터에 파일을 넘기기 전에 빠져있거나 서로 떨어져 있는 요소가 있으면, 출력시 모든 것이 무너질 수도 있었기 때문에, 그것을 하나하나 확인하는 작업이 마무리에서 가장 힘든 것이었는데, SEED TERMINAL의 나무 프레임들이 곡면에 또 그 갯수가 많아 서로 합쳐지는데 시간이 오래걸렸다. 무엇보다 주어진 절대시간 내에 출력할 수 있도록 해야했기에, 내부의 밀도를 조절하여 조금이라도 출력시간을 줄일 수 있도록 하는 것이 관건이었다. 하지만, 계획되었던 형태를 기존의 모델을 만들어오던 방식으로 했다면 전혀 표현할 수 없었던 곡선의 매끄러움이 3D 프린팅을 통해 잘 연출되었다는 것이 굉장히 만족스러웠고, 좋은 작업이었다.

MEGA SYSTEM

CHOI Ami KIM Jungae Matina CAVAYAS

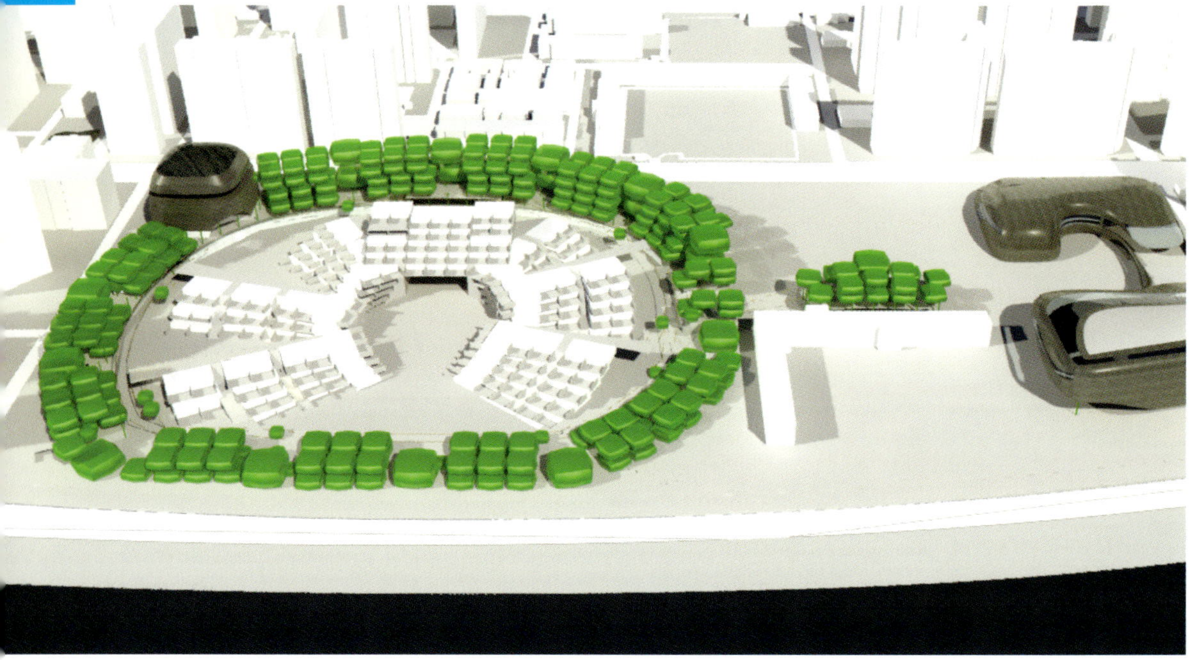

Perspective

Mobile Farm / Perspective

We chose 'healthy city, Seoul' as our design topic amongst the Bio-Digital City workshop topics. We suggested a methodology to provide every Seoulite with healthy food with a small design for a big change. Considering the 'food miles' of currently provided food in urban areas, most of it inevitably emits quite a large amount of carbon dioxide during the transportation and consumption process. Based on this observation, we designed the Mobile farm unit/Portable space which anyone can simply place at his/her house and move easily. The site was an apartment complex in Ichon. Starting at Ichon, our project suggests that Seoul develop towards a healthier city.

워크샵 주제인 Bio-Digital City, Seoul에서 우리가 디자인으로 발전한 주제는 '건강한도시, 서울'이다. 서울시민이 모두가 건강한 음식을 받기위한 방법론(작은 디자인 큰 변화)을 위한 디자인 아이디어를 제안하였다. 도심에서 제공되는 대부분의 식료품의 '푸드 마일'을 고려하여 볼 때 운송 과정 및 소비 과정에서 많은 이산화탄소(CO2) 를 배출하며, 결과적인 측면에서 운송비용이 실제 식품 가격보다 높게 설정 및 신선하지 못한 식품들이 많다. 각 가정에서 간단하게 설치 가능한 디자인 제품이나 이동가능한(Mobile farm unit / Potable space) 디자인을 고려, 제안하도록 한다. 사이트는 복합주거단지가 위치한 이촌 아파트지구로 지정하여 건강한 도시로서 발전할 수 있는 방안을 제안하고자 하는 프로젝트이다.

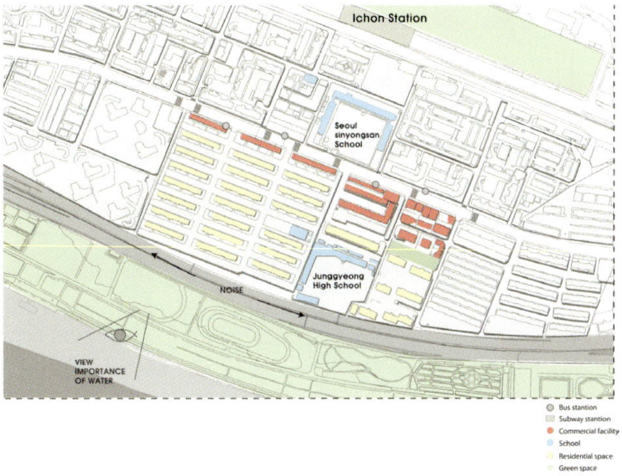

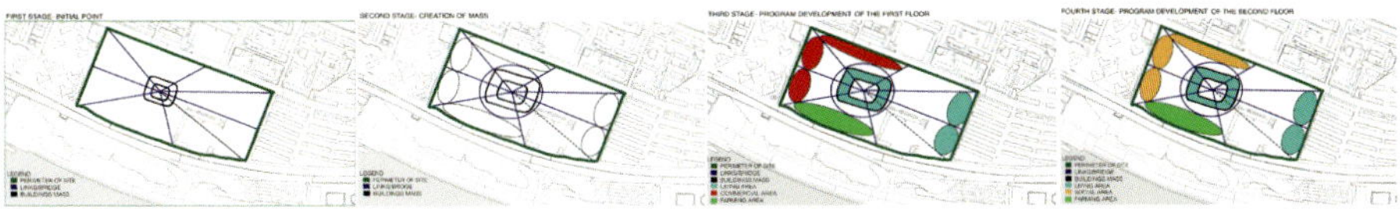

Spider Web Diagram

Bio Digital City Concept
Spider Web

In the South, Ichon Han River Park and Nodeul Island are located along the Han River and towards the North, the Yongsan Family park and National Museum of Korea are located. All the residential areas and educational institutions are located in between. The spot we chose to design is right in the middle between the National Museum of Korea and Ichon Han River Park. Within this area, apartment buildings, which are more than 20 stories, have been built to deal with the high population density along the Han riverside which resulted in the disconnection of parks and river views. There are 5 schools including elementary, middle and high schools within 2km as well as Joongkyung high school located inside of the site.
We focused on three points: the nature surrounding the area is disconnected, residential areas are located close to educational institutions, and plenty of leisure facilities are available. Then we designed a facility to connect these points without disturbing the current culture of the area.

서울시 용산구 이촌동은 남쪽에 위치한 한강을 따라 이촌 한강공원, 노들섬이 있고, 북쪽에는 용산가족공원과 국립중앙박물관이 위치한다. 그 사이 대지에 주거지역과 교육시설이 자리를 잡고 있다. 우리가 디자인하고자 하는 사이트는 국립중앙박물관과 이촌 한강공원을 잇는 정 중앙에 위치를 하고 있다. 현재 대지의 일대는 서울의 높은 인구밀도를 수용하기위해 한강변에 20층이 넘는 아파트를 세웠고, 그 결과 스카이라인이 높아져 공원과 한강의 조망이 끊겨있다. 그리고 대지 내를 포함하여 반경 2km 내에 초, 중, 고등학교가 5개가 있고, 그 중 중경고등학교가 대지 내부에 있다.
우리는 자연과 자연이 끊겨있다는 점, 주거단지와 교육시설이 공존한다는 점, 주변의 여가시설이 충분하다는 점에 초점을 맞추어 자연과 자연을 연결시키며, 현 대지가 가지고 있는 문화나 습관을 깨지 않고 어울릴 수 있는 시설을 구상하였다.

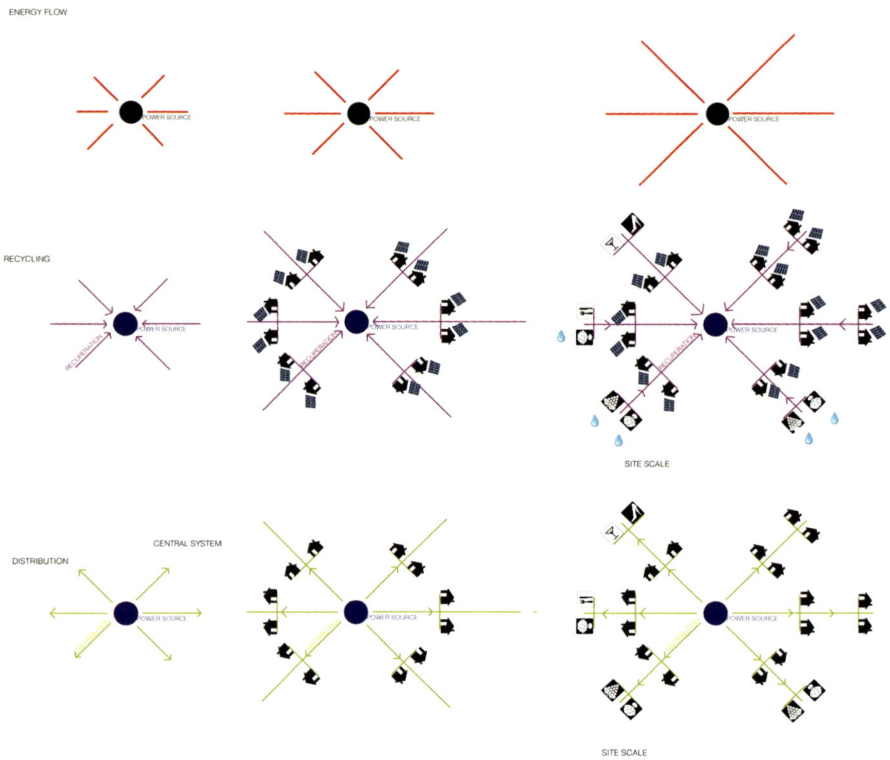

Design Concept

The final goal was to make the new village and the existing one compatible, so the school was to be maintained for Ichondong's character. Putting the school at the center, we planned two residential areas on both sides each having different characteristics. Considering the direction of the building towards the sunlight, a plant factory was located at the Eastern and Southern part of the site and commercial facilities were placed following the available accessibility. Thus commercial buildings were located at the Western and Northern part of the area closer to the main road while the plant farm was placed on the top of the northern commercial building. After sorting out functional frames, we placed a social plaza for interactions with residents from the outside as well as residents-only spaces.

최종 목표는 새로 만들어진 마을과 기존의 마을이 어울리는 것이었기 때문에, 이촌동의 분위기를 구성하는 학교는 그대로 두었다. 그래서 학교를 중심으로 양 옆에 서로 다른 성격의 주거단지를 계획하였다. 향을 고려하여 대지 내에 식물공장을 동쪽과 남쪽에 배치한 뒤, 접근성을 기준으로 상업시설을 배치하였다. 따라서 대로를 낀 서쪽과 북쪽에 상가가 위치하였고 북쪽의 상가 위에 식물농장을 얹었다. 전체적인 기능을 다 넣고 난 뒤, 외부 주민들과의 소통을 위해 사회적인 광장을 배치하고, 내부 주민들만의 공간을 따로 구성하였다.

Mega System

Mobile farm / Perspective

About 200 households of 120㎡ are arranged in a stair-case shape considering the angle of the light to form terrace houses. The distance between each building is thoughtfully adjusted to install rails so that small plant farms can move on it. Under the terrace house, amenities for residents were put in along with a plaza.

The plant farms surrounding the residential areas were placed in three units. 8 large (150x150x10M) farms, 120 middle sized (100x100x5M) farms, and 30 small farms that move amongst the large and middle sized farms. Because the farms are big enough, not only internal residents but also other Ichondong citizens can share and grow crops together, which enables more active interactions.

각 세대의 향을 고려해서 120㎡의 세대를 200여개를 계단식으로 배치하여 테라스 하우스를 만들었고, 간격을 조정하여 동을 만들어 그 사이에 레일을 설치해 소형 식물농장이 움직일 수 있게끔 하였다. 그리고 테라스하우스 아래 공간은 광장과 함께 내부 주민의 편의시설을 넣었다.

주거를 둘러싼 식물농장은 세가지 유닛으로 나누어 배치하였다. 150X150X10M의 대형 농장이 여덟개, 100X100X5M의 중형 농장이 120개, 중형과 대형 농장 사이를 움직이는 소형 농장이 30개가 들어갔다. 식물농장의 크기가 크므로 내부 주민의 텃밭만이 아닌 외부의 이촌동 주민과 함께 가꿀 수 있는 텃밭으로 하였다. 이로인해 이촌동 주민들 사이에 만날 수 있는 공간은 늘어나게 되었다.

1. 학교를 중심으로 사이트를 이등분 하여 주거시설을 배치한다.
2. 채광과 접근성을 고려하여 상업시설과 식물공장을 배치한다.
3. 사이트 외부 주민과 내부 주민을 위한 광장을 구성한다.

1. Place residential buildings by dividing the site into two with the school at the center.
2. Place commercial facilities and plant factory with light and accessibility considered.
3. Build a plaza for residents and visitor from outside of the site.

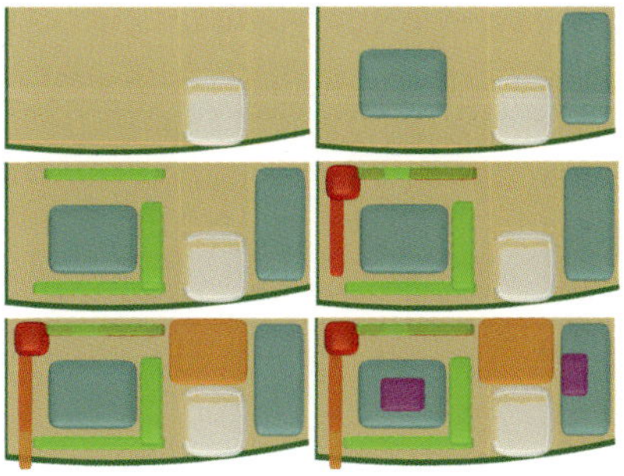

1

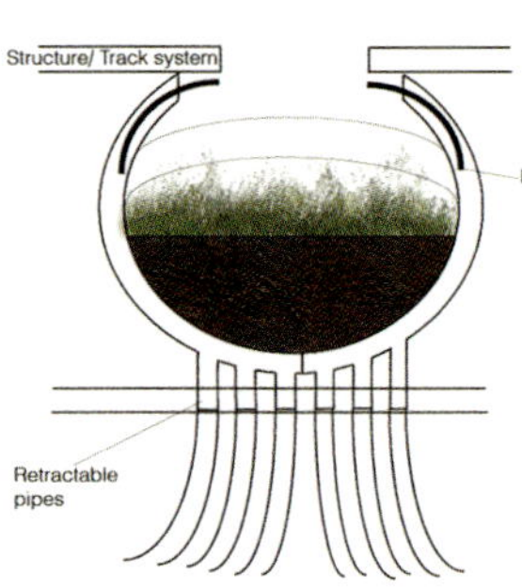

2

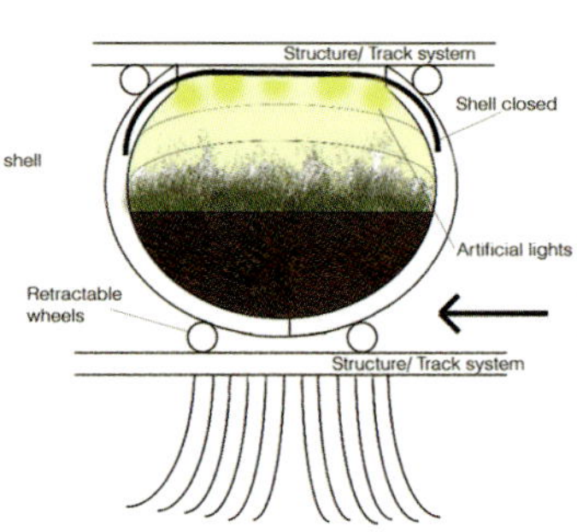

Farm Unit

Farm Unit Detail / Fram Unit Work

Small plant farms move around to residential areas and to the other plant farm complexes. The biggest issue to solve was the systematic management of the soil and drainage. We designed flexible holes at the bottom which can be connected to tubes when it stops so that soil and drainage can be managed. When they moving, the holes are closed.
The external layer of the plant farm consists of spider web fiber. It can contract and expand according to the strength of the light so the illumination inside the farm can be controlled. The amount of rain is also calculated so when the rain reaches a certain point, the external layer closes to store the extra rainwater. The layer exists in our imagination for now, but once it comes true, the energy of nature can be utilized in innovative ways.

The residential areas in the East have different characteristics compared to the terrace houses because it has less contact with the plant farms. Therefore, they are built in more flexible forms so that the shape itself can be distinctive. We limited the height to 7 stories tall so the view from the park toward the Han River is undisturbed.

소형 식물공장의 경우, 주거단지와 식물공장군 사이를 다닌다. 가장 큰 논의는 흙의 체계적 관리와 배수의 처리이다. 따라서 바닥면에 열었다 닫았다 할 수 있는 유연한 구멍을 만들어 정지하였을 때는 구멍에 관을 연결하여 배수와 흙의 교체의 용도로 사용하고, 움직일 때는 구멍이 닫혀 바퀴로 이동이 가능하게끔 구상하였다.
식물공장의 외피는 거미섬유로 이루어져 있다. 이 거미섬유는 빛의 강도에 따라 수축과 팽창을 하여 식물 공장 내 조도를 조절하기도 하고, 비가 오면 공장에 떨어지는 물의 양을 계산하여 일정량 이상이 될 때 외피를 닫아 잉여 빗물을 저장한다. 현재는 우리의 상상 속에 존재하는 외피지만 실현된다면, 사라지는 자연 에너지를 보다 획기적으로 이용할 수 있을 것이다.

동쪽의 주거시설은 테라스 하우스와 달리 식물공장과의 접촉이 적어 성격이 달라졌다. 따라서 형태에서부터 먼저 구분을 두어 폼을 보다 유연하게 만들어보았다. 높이는 7층 정도로 두어 공원에서 한강을 바라볼 때 시선이 깨지지 않도록 하였다.

3D Print Model

Farm / Home
Commercial / Spcial

We were a Canadian from Quebec and two Koreans. The language barrier was an issue at first but after spending 15 hours a day together during the workshop we began to understand how to communicate with each other. We came up with the metaphor of a Spiderweb fiber by connecting Ichondong's historical features and Bio digital together. As a result, we tried to bring natural spaces into the site and to also create a way to make coexistence continue.
We experienced a lot of errors with the 3D printing, much more than any other team. The plant factory developed into the form of a cell so the machine couldn't recognize it, not to mention that we almost failed to submit due to certain empty spots in the middle of our model. After several modifications, our results were a bit different from our initial concepts which we interpreted as a part of the technical studies required.

우리 조는 퀘백 출신의 캐나다인 한명, 한국인 두명으로 이루어져 작업을 하였다. 처음엔 언어의 벽에서 서로 의사소통이 잘 되지 않아 작업에 난항을 겪었지만, 워크샵 기간 동안 하루 15시간을 함께 하면서 서로를 이해하는 방법을 터득하였다. 그 결과 이촌동의 역사성과 바이오디지털을 연결시켜 거미섬유라는 메타포를 생각해내었다. 목표로 했던 기존의 단절된 자연공간을 연결시켰고, 그 속에서 공존하는 방법을 만들어냈다.
3D 프린터를 하는데, 우리 조는 다른 조와 다르게 상당히 많은 오류가 났다. 애초 식물공장이 세포의 형태가 되어 기계가 인식하지 못한 것도 있었고, 중간 중간 비어지는 부분 때문에 프린트 결과물을 제출하지 못할 뻔하기도 하였다. 몇 번의 수정 끝에 초안과는 다른 모양이 나오게 되었지만, 이 과정 역시 기술을 이해하는 공부의 일부분이었다.

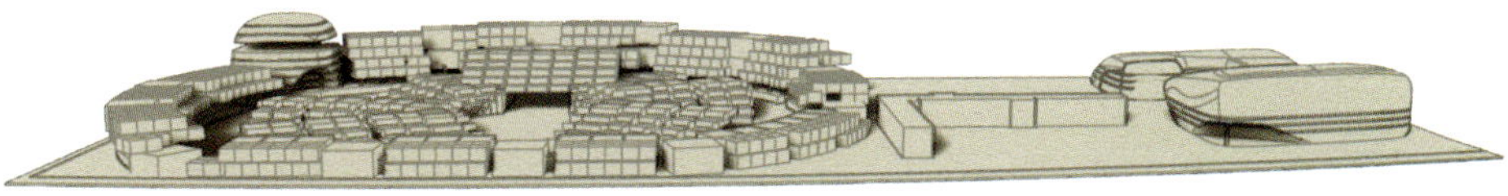

PROFILES

KIM, Guyhong

Born in Busan, Korea. Bachelor's degree from Kookmin University, Department of Architecture and from University Paris 8 Department of Plastic Arts. Master of Architecture from Paris-La Villette School of Architecture.
Architect DPLG, CEO of Urban Design KyOsK.
Korean Traditional Space photo exhibition in Lille. Lecturing about 'Media City' for the HCI Association. Selected amongst 100 Young Designers by ANC. Invited Architect, KIA.

LEE, Jongeun

Graduated from Kookmin University, Department of Architecture. Master of Architecture from Paris-La Villette School of Architecture.
Architect DPLG.
Worked at Mooyoung Architectural Firm on apartment housing and methodology to improve residential environment of Korea.
Operates 3NA Neuro-Architecture LABs since 2013, striving to create space that meets the various needs of occupants with scientific data on the physiological and psychological influence of physical space on its users.

JANG, Suho

Born in Suncheon, Korea. Bachelor and Master's degree from Suncheon National University, Department of Image Design.
PhD candidate and researcher at the Digital Art Technology Application Lab, Chung-ang University. CEO of Hyperstage

MOON, Junsik

PhD candidate at Seoul National University Architecture Department.
Teaches courses on the fundamental theories of computer graphics relevant to design at Konkuk, Yonsei and Inha Universities.
His research focuses on the representation and computation of architectural space, interactive architecture and digital architecture.

KO, Suejeong

Born in Busan, Korea. Bachelor's degree from Pukyong National University, Department of Architecture. Master's degree from Hongik University School of Architecture and University of Pennsylvania.
LEED AP BD+C. Professor at Pukyong National University.

KIM, Dongjo

Bachelor's degree from Suncheon National University, Department of Image Design. Master and PhD degree in graphic engineering – art engineering from Chun-ang University.
Artist in residence at Geumcheon Art Camp. Part-time lecturer at Incheon Catholic University and Suncheon University.

SONG, Sun

Attended Ravensbourne College of Design and Communication, London. Bachelor and Master's degree from Chung-ang University, Multimedia and Film. Art & Tech PhD from Chung-ang University.

LEE, Youngseok

Bachelor's degree from Yonsei University. Master's degree fromm AA School and Cambridge University.
Professor at Kyunghee University, Department of Architecture. CEO of Urbanindest Lab.

KIM, Seokwon

Bachelor and Master's degree from Seoul National University Architecture Department. Won a Scholarship from the Korean Government. Master of Architecture's Degree from Bartlett, UCL, directed by Sir Peter Cook. Master's degree in Histories and Theories from the AA School of Architecture.
Worked in the Urban Division and Architectural Design Department of SPACE Group.
PhD candidate at Seoul National University. Teaching at SNU, Hanyang University, Sejong University and Suwon University.

WOO, Junghyun

Bachelor's degree in Environment Design from Ewha Woman University. Master's degree from the Berlage Institute.
Worked at MVRDV Rotterdam, The Relational Urbanism Office, London and SIAPLAN Seoul and New York. Since 2012, lectured in various universities in South Korea, the USA and the Netherlands.
Research assistant for the EXUMA Lab directed by Professor Charles Waldheim, Chair of the Department of Landscape Architecture at Harvard Graduate School of Design.

AN, Na

Bachelor's degree from Seoul National University of Science and Technology, Art Major. Studied Space Design at Chelsea College of Art an Design. Master's degree of Architecture and Urban Design from the Berlage Institute.
Works at ANKUNDESIGN GROUP, Korea. Member of the Korean Institute of Culture Architecture.

JUNG, Hyunyi

Bachelor's degree in Dance from Rutgers, State University of New Jersey. Master's degree in Art & Technology from Chung-ang University.
PhD candidate at the Graduate School of Advanced Imaging Science, Multimedia & Film, Chung-ang University. Lecturing in universities and planning New Media Art.
Her research interest include media performance, interactive art and projection mapping.

SEARCHING FOR A NEW ARCHITECTURAL AND URBAN MODERNITY

Jean MAGERAND Claire BAILLY

L'obsolescence de l'héritage moderne

Après la Modernité salvatrice de l'après deuxième guerre mondiale, qui a fait rêver pays riches et pays pauvres, c'est une Modernité nocive, pétrie de désillusions, de mauvaise conscience qui a, peu à peu, imposé sa présence dans tous les pays. L'Homme nouveau, ainsi que son jumeau l'Homme moderne, inventés par l'avant-garde du 20ème siècle ont été adulés jusque dans les années 1960. Ils sont aujourd'hui totalement remis en cause. L'Homme Standard de Le Corbusier, icône du progrès, s'est révélé être un gaspilleur égoïste qui détruit son environnement. Il néglige ses semblables, les écosystèmes et les générations futures.
Dans ce contexte de «mauvaise modernité» et particulièrement au cours des dernières décennies, la finalité même des connaissances objectives a été remise en cause. Les nobles sciences, les techniques et leurs enfants naturels, les techno-sciences, brandies pendant un siècle comme machines à conquérir la nature, comme génératrices de progrès et de bonheur, comme salvatrices de l'humanité, ont peu à peu été reléguées au statut de « complices de la détérioration de l'environnement ». L'introduction d'une dose de « retour aux valeurs anciennes » portée par un « historicisme politiquement-correcte » le tout complété par un zeste de progressisme a semblé être, pour toute une génération, le remède idéale contre les dérives modernistes. A cela, il faut ajouter le grouillement de remises en causes et de questionnement, généré par certains phénomènes et événements, déterminants pour le contexte contemporain. Ainsi, le changement climatique est considéré comme résultant de la techno-économie mondialiste qui est elle-même accusée de folie destructrice voire auto-destructrice. Mais encore, la révolution biologique induit la naissance d'une bio-éthique qui ne fait que compliquer les interrogations.
Dans le même temps, les savants, sachants et autres experts ont découvert les tenants et aboutissants des sciences de l'information et du temps réel ; ils ont pris conscience des conséquences techniques et culturelles de cette nouvelle donne. Ils ont très vite compris que de nouvelles méthodes, apparues dans ce contexte, sont en train d'influencer toutes les sociétés ; elles accélèrent les phénomènes de mondialisation qui imprègnent tant les domaines de l'économie que ceux de la culture. Tout changement brusque induit aussitôt de fortes réticences. Notamment, la maîtrise de l'information et sa servante, l'informatique, ont été suspectées d'avoir une mauvaise influence sur les individus et sur leurs collectifs. En plus de cela, les crises économiques sèment la peur dans les populations et menacent en permanence, quand elles ne sont pas effectives. Cette ambiance de mise en doute crée une « sur-pression » qui induit aussitôt des effets de «dépression perceptive» modifiant le regard que nous portons sur notre environnement, sur notre culture, sur nos habitudes, sur notre modernité. Au sein de cette mutation de nos icônes et valeurs sociétales, c'est subitement un regard inversé qui est porté sur le rapport Homme/milieu. La planète incommensurable il y a quelques décennies, a pris les dimensions d'un bocal.
Dans ce contexte de crise des valeurs, de changement contraint de notre environnement biologique, de mutations scientifiques et techniques mal digérées, se crée un climat de tensions troubles qui rend flous les enjeux du débat en camouflant les causes et leurs conséquences. La compréhension de ces phénomènes en marche n'est pas simple car la mutation du contexte - enserrée dans un univers technique inhabituel - génère une technophobie et une technophilie qui s'affrontent en un duel/débat tout à la fois passionné, doctrinal, superficiel et plein d'incompréhension. Les postures critiques sont remplacées par des positions de principe. Nos névroses collectives perturbent donc les regards qui se veulent pourtant objectifs et/ou distanciés sur notre monde contemporain. Tout ce fatras d'a priori et de discours contradictoires rend difficile une appréciation de l'«air du temps» et par là même un jugement objectif sur le bien fondé des pensées conceptualisantes de notre époque.
Notre objectif dans ce monde en questionnement est de chercher un recul suffisant afin de mieux discerner les incidences de ce contexte complexe sur l'évolution des pensées, discours, théories, pratiques et productions liées à l'aménagement et au cadre de vie.
Comme repère de cette situation ambiguë, nous avons l'histoire, celle qui s'intéresse aux mécanismes, pas à la seule symbolique des faits. Cette histoire nous apprend que cette situation de grand questionnement est récurrente dans l'évolution des cultures et des civilisations. A certaines époques charnières, des discours totalement contradictoires co-existent de manière forte. Le bilan a posteriori montre que des signes permanents habitent ces époques particulières et qu'il est possible d'y retrouver des éléments de constance susceptibles de nous éclairer sur notre présent. Nous sommes à la recherche de ces signes et de leurs propriétés. La révolution numérique et la révolution de la connaissance des mécanismes du vivant nous plongent au sein d'une de ces périodes charnières. C'est de là que pour nous émerge le concept de Bio-numérique.

Les gènes d'une culture mutante

Le rejet des techno-sciences est bien évidemment très loin d'être suffisant à lui seul pour reformater l'«air du temps» ou même le sens de l'histoire. Dans ce contexte de « modernité déclinante », l'Agriculture, la Ville, la révolution informatique sont considérées, dans les discours dominants, tout au plus comme les responsables de la destruction de la planète et comme la cause du mal-être mental et physique de la société. Ces doctrines procèdent en fait d'une logique d'empêchement et de frein, mais elles ne sont aucunement porteuses d'une dynamique de changement raisonné. Il est donc très important de bien discerner ce qui apparaît et ce qui disparaît dans notre civilisation contemporaine.

Tout montre qu'aujourd'hui il est également nécessaire, afin d'éviter à terme l'auto-destruction de la planète et de préserver la vie sur terre, de mieux mettre en évidence les pistes potentielles d'une nouvelle construction civilisationnelle. Pour mieux connaître notre époque, il nous faut bien décrire et évaluer les agents actifs des mutations en cours.

Par exemple, actuellement, les phénomènes d'évolution des modes d'analyse du réel, fondés sur des modèles écosystémiques, sont amplifiés par les nouveaux outils de lecture issus des mutations méthodico-techniques. En se constituant ainsi, le nouveau contexte de vie et de pensée agit directement sur notre manière d'analyser le réel et donc de l'aborder et de le comprendre.

Aujourd'hui, une réaction en chaîne se met en route, générant les hyper-lectures artificielles d'un réel sur-complexifié. Notre milieu artificiel ainsi re-présenté et donc re-constitué change notre manière de le percevoir et de nous percevoir, en son sein. Se pose aussitôt la question des conséquences diverses de ces nouvelles lectures sur nos manières de ré-écrire du réel, d'en créer. Cette même question se pose pour toutes sortes de mécanismes, dispositifs, processus nouveaux qui se mettent en place dans le sillage des nouvelles techniques et de leurs univers méthodiques inédits.

C'est une fois mieux éclairés sur ces aspects nouveaux de notre époque que nous serons à même de nous forger les armes utiles pour mieux comprendre les mécanismes qui encadrent l'influence des récentes approches scientifiques du réel sur d'autres domaines tels ceux de la conception.

Ce fameux air du temps, que nous cherchons à introduire dans la conception, les promoteurs du Mouvement moderne, par exemple, le considéraient comme un «formatant inéluctable». Il dépendait totalement, à leurs yeux, d'une multitude de réels théoriques, conceptuels, matériels, immatériels, artefactuels ou naturels constituant les avant-gardes. Pour eux, toutes ces spécificités hétérogènes, orientées par la boussole d'un « progrès » parfaitement identifiable, s'associaient et s'hybridaient en générant un contexte commun de pensée et de vie qui lui conférait sa substance.

Aujourd'hui, les travaux d'Edgar Morin (La nature de la nature) ou de Joël de Rosnay (L'Homme symbiotique) nous suggèrent un air du temps « profond » où les unités, multitudes, les multiplicités, les systèmes, les appareillages, les certitudes, les ambiguïtés s'enchevêtrent en une complexité rationnelle, mais inaccessible dans sa cohérence globale. Ils nous ouvrent à des univers évocables mais insondables dans leur entièreté. Dans ces nouvelles philosophies, ces réels constituants, imbriqués les uns dans les autres et composant le réel majeur de notre milieu complexe, sont approchés sous un angle global spécifique. La finalité de ces nouvelles re-compositions du réel est de mieux mettre en évidence les organisations de ces réels coexistants et de mieux comprendre les mécanismes régissant leurs activités, leurs rapports, leurs évolutions et leurs capacités à agir.

Aujourd'hui, quelques grands phénomènes paradigmatiques révolutionnaires structurent le contexte contemporain ; ils semblent très prometteurs en termes de changement. Ainsi, les sciences de l'information ont engagé une mutation technique qui à son tour provoque, directement et indirectement, des révolutions comportementales, du fait, entre autres, du caractère exponentiel de la circulation de l'information. L'installation du temps réel, chaque jour plus présent et plus pressant, permet en fait de lutter contre les effets pervers des phénomènes de mondialisation qu'il a pourtant contribué à générer. Les Sciences de la complexité aiguisent et amplifient notre compréhension du réel en générant, in fine, une révolution intellectuelle. La conscience nouvelle de nos environnements - boostée par les sciences de la complexité - contribue aussi à re-formater et à faire muter nos regards sur l'environnement. Nous sommes en mesure d'envisager autrement nos actions au sein de l'écosystème planétaire. La biologie est de plus porteuse d'enjeux primordiaux, elle surgit au devant de la scène et propose de nouvelles approches du vivant toujours moins magiques mais toujours plus époustouflantes. Elle est certes porteuse de grandes peurs mais aussi d'espoirs immenses.
Le numérique s'infiltre dans la production des connaissances du réel en générant les mêmes angoisses et les mêmes fascinations. Une conscience nouvelle et une révolution culturelle, à l'échelle de toute la planète, d'une amplitude sans équivalent dans l'histoire, semblent lentement émerger. Toutes ces nouvelles données, formant entités civilisationnelles, s'agglomèrent et/ou se mixent et/ou se mettent en synergie pour influencer l'évolution de l'humanité et de ses savoirs. C'est ensemble, qu'elles enclenchent cette mutation majeure de notre rapport à nos environnements sociétaux, culturels, sociaux, conceptuels. A notre époque, comme à toutes les autres périodes charnières déjà vécues par l'humanité, la nature et les positions relatives de chacune de ces variables humaines varient au sein de ce contexte. Le regard qui est porté sur elles évolue également. Comme dans tout système, lorsqu'il est abordé sous l'angle de sa structure écosystémique, le contexte formate chacune de ses sous-parties. Un dispositif d'hybridation plus ou moins actif et effectif se met en place selon les lieux et le moment. En retour chaque entité constitutive réagit en synergie ou en contradiction avec les tendances générales ; elle participe au changement du contexte. Ainsi contrainte par le système, elle évolue par analogie, réaction, inertie ou rétroaction ; elle mute en s'adaptant. C'est dans ce contexte très nouveau et très particulier que nos modes de vie et nos cultures sont susceptibles de changer en même temps que nos exigences et/ou nos besoins de nouveaux cadres-de-vie mieux adaptés aux nouvelles situations que nous nous sommes nous-mêmes imposés.

A la recherche d'un cadre de vie pour une Nouvelle modernité

C'est conscient de ces réticences contemporaines et de la morphologie complexe de ces nouvelles mutations civilisationnelles et philosophiques que l'Atelier international expérimental pour la cité bio-numérique poursuit ses recherches, depuis plusieurs années, sur les conséquences de cette « nouvelle donne » sur nos milieux de vie, c'est-à-dire sur notre cadre de vie planétaire. Quels seront nos territoires naturels, nos paysages, nos agro-systèmes, nos architectures, nos villes, nos sociétés demain? Comment habiterons-nous, travaillerons-nous, collaborerons-nous? Comment pourrons-nous revoir nos manières de faire pour ne plus gaspiller le territoire, les richesses naturelles, le patrimoine? Comment pourrons-nous abandonner la mondialisation suicidaire fondée sur la concurrence, l'antagonisme, la compétition? Comment pourrons-nous mettre en place des sociétés mieux civilisées, moins égoïstes, moins gaspilleuses, plus solidaires, plus tolérantes, plus collaboratives, plus concertantes, plus fondées sur l'entre-aide, le partenariat, le partage, la solidarité, la mutualisation? Il est bien évident que ces qualités ne sont plus seulement porteuses de leur seule valeur morale mais qu'elles constituent une clé possible pour inverser le lent processus auto-destructif dans lequel les organisations humaines se sont aventurées depuis un siècle et demi.
L'une de nos hypothèses de travail est que cette révolution culturelle, incluant les valeurs morales, est un passage obligé, au sein duquel l'outil numérique occupe un positionnement stratégique. Il est en effet devenu un dénominateur commun entre d'une part le grand public utilisant massivement le numérique au quotidien et d'autre part l'Entreprise qui est contrainte, pour son développement, d'acquérir chaque jour des compétences nouvelles en la matière. De notre point de vue, ce grand public n'est pas encore convaincu mais il est déjà sensibilisé aux travaux de recherche high-tech qui visent à des améliorations environnementales à l'échelle de l'appartement(domotique), de l'immeuble(immotique), du quartier(smart-city), du pays(transports intelligents) comme à l'échelle de la planète(développement durable assisté par satellite). Ces techniques peuvent s'appliquer aux territoires urbains ou agricoles ou bien encore s'hybrider avec d'autres savoir-faire existants pour constituer de nouvelles techniques d'amélioration des conditions de vie en ville et à la campagne.

Vers une gestion complexe des territoires

Nos recherches se penchent sur une agriculture nouvelle, biologique et numérique. Hydroponie, permaculture, agro-foresterie s'hybrident avec la robotique afin d'augmenter les rendements sans polluer les sols et sans porter atteinte aux équilibres naturels.
Cette agriculture est entièrement automatisée, elle est optimisée et à haute performance. Les champs sont cultivés par de petits robots qui marchent à l'énergie solaire. Ils entretiennent les sols, ils arrosent, désherbent mécaniquement, taillent les végétaux, coupent l'herbe, broient, cueillent à l'unité, stockent, suppriment les insectes nuisibles, font fuir les oiseaux, mettent la production en sac ou en boite, transportent. Ils participent également à la gestion et à l'entretien des fermes d'élevage. L'Agriculture auto-organisée produit des fruits et légumes, des agrocarburants et du méthane mais elle recycle aussi les déchets urbains biologiques. Elle épure l'eau, favorise la biodiversité, crée des paysages, des lieux de promenades et des zones de jardinages pour les familles. Les robots travaillent jour et nuit, silencieusement. Ils entretiennent les arbres de l'agro-foresterie et maintiennent un ensoleillement optimisé. Les lombricultures robotisées produisent du compost, des engrais naturels liquides, et de la nourriture pour les poissons des piscicultures. L'eau est épurée par des phyto-bassins automatisés qui produisent des plantes comestibles ou des algues utiles pour produire des engrais verts ou des carburants. Les fermes génèrent les paysages diversifiés à partir de bocages de petits champs avec des cultures mélangées. Les fermes-urbaines-automatiques engrangent des données et alimentent les Big-Data en transférant leurs informations aux satellites. Ces équipements agronomiques s'hybrident aux milieux naturels et aux milieux urbains en les bonifiant. Mais l'agriculture n'est pas seulement une immense machinerie, elle doit aussi être productrice d'un univers méthodique et culturel valorisant pour l'homme et elle doit faire vivre correctement ses agriculteurs.
L'architecture des hypercités est aussi une architecture adaptée à une robotisation intensive. La majorité des effluents et des déchets est recyclée à l'intérieur même de l'immeuble. A coté de cette agriculture high-tech, la ville est dessinée pour que les robots créent le trait d'union ville-campagne en assurant aussi les recyclages courts. La nuit, toute la ville est entretenue avec les mêmes outils automatiques, tout aussi silencieusement que dans les champs. Lorsque les métros ne circulent pas, les tunnels entre stations sont utilisés pour trier les marchandises. Les déchets sont comprimés et mis en containers étanches. Des wagons de marchandise circulent sur les voies. Ils apportent les aliments frais directement dans les différents quartiers. Ils sont chargés et déchargés par des robots. Les déchets biologiques sont transportés par rails jusque dans les zones agricoles ou dans les zones de compostage. En étant automatisés selon les principes de l'auto-organisation, la ville génère un milieu propice à la vie humaine. Loin d'être un environnement purement technique, la ville et l'architecture se fondent sur le principe des phénomènes émergents. Elles sont sensibles à la dimension et à la condition humaine. Elles s'adaptent en temps réel aux contraintes, aux priorités, aux impératifs, aux besoins, aux nécessités d'une société mieux civilisée, plus tolérante, plus transparente.
Il s'agit de prendre en compte des technologies déjà au point et en train de se mettre en place dans nos sociétés contemporaines. Toutes les nouvelles techniques et nouvelles méthodes doivent être au centre des préoccupations de durabilité. Elles favorisent le confort, le bien-être, le mieux vivre, la convivialité, la tolérance, la collaborativité.
Les prototypes de cité numérique sont donc des « Hyper-villes » à très haute qualité de vie. Ils sont denses, économes en énergie, en eau et en ressources naturelles. Ils concentrent, exposent, améliorent et valorisent des savoir-faire technologiques sophistiqués (traitement robotique des déchets, recyclage intelligent de l'eau, robotique agricole, stockage urbain robotisé, fabrication à distance, restaurants bio-robotisés et bien d'autres encore). Ces villes contibuent à ré-équilibrer en temps réel les écosystèmes planétaires et les territoires agricoles. Elles intègrent une agriculture urbaine biologique et high-tech.
Les fab-lab/usine en milieu urbain et ruraux sont développés afin de favoriser la lutte contre les gaspillages énergétiques et des matières premières. Les organisations industrielles et économiques sont re-visitées en exploitant au maximum les potentiels des nouveaux outils de gestion et d'information et en favorisant les multi-recyclages courts.

Nous considérons que, plus largement, tous les domaines techniques ou industriels applicables de près ou de loin au territoire sont concernés par cette mutation de la production/ consommation. Par exemple, matériels agricoles robotisés, aménagement environnementaux, robotiques de chantier, assainissements intelligents, tris sélectifs automatisés, recyclages paramétrés, transports expérimentaux, stockages informatisés, environnement adaptatifs, véhicules spéciaux, habitats domotisés sont des domaines appelés à renouveler les modes de production industrielle.

Ces recherches interrogent tous les savoirs et chacun d'entre nous, à titre privé, sur la mutation que nous vivons aujourd'hui, qu'elle soit culturelle, civilisationnelle, intellectuelle, économique industrielle ou environnementale. Dans cette ambiance technique et méthodique, est évidemment présent le sujet majeur de l'invention de nouvelles formes de collaboration entre les individus, les collectivités, les sociétés, les pays. Les traumatismes laissés par les utopies techno-scientifiques de la premières moitié du 20ème siècle demeurent encore. Il est donc utile de manipuler avec beaucoup de précautions les propositions d'un nouveau cadre de vie lorsque ce dernier prétend s'inspirer de nouvelles organisations techniques et scientifiques.

L'hypothèse-mère de ces recherches est donc que la dimension organisationnelle est transversale, qu'elle est solidaire et interdépendante de la dimension sociétale et qu'elle peut s'enrichir de la « substance » informatique, indépendamment des aspects purement techniques de cette dernière. Dans ce contexte, la question est donc posée, de savoir comment des outils rationnels peuvent inspirer des collectifs plus solidaires, plus tolérants, plus humanistes sans forcément engager nos sociétés dans des dispositifs autoritaires confisqués par quelques-uns.

L'Atelier international expérimental pour la Cité Bio-numérique, un outil expérimental et hybride

L'Atelier international expérimental pour la Cité bio-numérique s'est fixé pour mission de développer ses recherches prospectives et de sensibiliser le grand public aux nouveaux enjeux liés au numérique . Dans cette optique, une place importante est laissée à la réflexion sur les solutions matérielles ou immatérielles, qui permettront à terme de contribuer à mettre fin aux grands désordres provoqués par le développement des villes et les agricultures nocives pour la planète. En outre, l'Atelier souhaite faciliter, générer et/ou solidifier des collaborations techniques et scientifiques nouvelles afin de déboucher sur des inventions opérationnelles et par là même générer de nouvelles dynamiques économiques.

En Août 2013, la Cité des Sciences et de l'industrie de Paris accueillait l'Atelier international au sein même de ses expositions prestigieuses. Ainsi, le grand public a pu se familiariser avec de nouvelles approches de la ville de demain. C'est dans ce contexte que l'Atelier produit et valorise ses recherches en France, en Belgique et en Corée.

L'Atelier produit, grâce au numérique, des prototypes urbains auto-organisés et intelligents, des modèles de cités hyper-recyclantes, des architectures adaptables en temps réel, des agricultures bio-optimisées, des organisations individuelles et collectives plus solidaires, mieux mutualisées, tous bénéfiques pour la planète.

Il s'agit d'identifier, utiliser, valoriser et croiser, dans les projets urbains et ruraux, les connaissances techniques, scientifiques, industrielles, sociologiques les plus pointues. A partir de nouveaux matériaux, de nouvelles techniques, de nouveaux savoir-faire, de nouveaux arts de vivre, l'objectif est de concevoir, de construire et de gérer les cités par des méthodes optimisantes et innovantes (algorithmie, modélisation, optimisation paramétrée, travail collaboratif etc.).

L'Atelier est permanent. Il associe, sur dossier, les jeunes spécialistes les mieux rodés aux nouveaux outils, au sein d'une formule collaborative souple ; ils sont architectes, urbanistes, paysagistes, artistes, ingénieurs, informaticiens, agronomes, géographes, mathématiciens, roboticiens, sociologues, historiens, biologistes, écologues ou tous autres spécialistes agissant sur les territoires de la Cité bio-numérique.

L'Atelier travaille autour de plateformes collaboratives afin de mettre au point des méthodes d'expérimentations conceptuelles intégrant les recyclages intégraux ultra-courts, la robotique agricole et de chantier, la biodiversité assistée, l'habitat à haute valeur technique et environnement, la sociologie numérique etc.

Le workshop bio-numérique 2014 à la Mairie de Séoul, une étape importante dans la recherche de modes de conception hybrides

Le workshop organisé à Séoul s'inscrit dans la suite conceptuelle des Wokshops organisés en collaboration avec la Faculté d'Architecture de la Yeungnam University en Corée du sud, le laboratoire EVCAU (Environnements virtuels et Conception Architecturale et Urbaine), l'école d'architecture de Montpellier ou la Faculté d'Architecture de Mons en Belgique. Les workshops se déroulent au Palais de Tokyo, à la Cité des Sciences à Paris, à Mons durant « Mons 2015 /Capitale européenne de la Culture », à l'hotel de ville de Séoul. Les partenaires habituels sont le ministère français des affaires étrangères, Universcience à Paris, l'Institut Français, l'Institut français en Corée du sud, l'Association des architectes DPLG / HMONP de Corée, l'IUI (International Urbanism Institut), l'Atelier international du Grand Paris. Des recherches, projets et expositions sont prévus en Corée et en France en 2014, 2015, 2016 (Paris, Mons, Séoul). L'Atelier établit des contacts privilégiés avec les grandes écoles, les universités, les administrations, les industries de pointe, les collectivités territoriales.
Le workshop regroupe des jeunes spécialistes. Ils rassemblent des informations et des actions sur les modes d'enseignement de l'Architecture, sur les technologies les plus avancées, sur les modes de vie du futur, sur la maîtrise de l'énergie, sur le rapport ville/changement climatique, sur les nouvelles manières de penser et de dessiner les villes de demain. Les participants (enseignants et étudiants) sont amenés à réfléchir ensemble aux nouveaux modes de vie, aux nouveaux « modes de ville », aux nouvelles agricultures, aux nouveaux déplacements, aux nouvelles productions/consommations tous rendus possibles par l'univers scientifique et numérique contemporain.

BIOGRAPHIE

Au sein de l'« Atelier international expérimental pour la Cité bio-numérique », Claire Bailly et Jean Magerand organisent la production architecturale, urbaine et paysagère ainsi que les travaux de recherche, et la pédagogie. Ils sont architectes, urbanistes, paysagistes. L'une est titulaire d'un DEA en en science de l'information, l'autre titulaire d'un doctorat en architecture. Ils ont effectué de nombreuses publications. Leurs projets ont été primés à l'occasion de plusieurs concours internationaux d'architecture de paysage et d'urbanisme sur la ville du futur.

Marlène VERNOUILLET, Scientific facilitator
At the Cité des sciences et de l'industrie, Universcience, Paris

THE ROLE OF FACILITATORS IN WORKSHOP PROGRAMMES

Avant-propos

J'ai participé au workshop de Séoul du 5 au 14 août 2014 dans la continuité des workshops qui ont eu lieu à la Cité des sciences en 2013. Ce workshop s'inscrit pour la Cité des sciences dans un dispositif large qui a notamment pour objectif d'intégrer le citoyen dans une réflexion concernant l'habitat urbain et le milieu rural.

Un constat : le modèle urbain en rupture

L'urgence est là, et pour tous, architectes, urbanistes, designers, habitants, de repenser la ville de demain sans détruire les terres agricoles. « Ré-inventer nos lieux de vie » est une nécessité pour notre futur car les modèles urbains contiennent de plus en plus de failles et sur de nombreux aspects.
D'un point de vue humain, la population mondiale augmente et donc la population urbaine. Depuis 2007, la population urbaine a dépassé la population rurale. En 2030, d'après les estimations, deux tiers de la population mondiale vivra en ville.
De plus, on observe également un vieillissement de la population. A l'heure actuelle, il y a plus de 606 millions de personnes de plus de 60 ans et 379 millions de plus de 80 ans. D'ici 2050, le nombre de plus de 60 ans sera multiplié par 5 et celui des plus de 80 ans par trois. Quels seront les aménagements urbains nécessaires pour permettre à cette population ayant des besoins spécifiques notamment lié à la perte de mobilité ?
Par ailleurs, on constate une augmentation des inégalités sociales. Le phénomène de gentrification (des arrivants plus aisés s'approprient un espace initialement occupé par des habitants ou usagers moins favorisés, transformant ainsi le profil économique et social du quartier au profit exclusif d'une couche sociale supérieure) s'intensifie. Les gens vivent dans l'anonymat et sont de plus en plus isolés. Les lieux de vie collectifs sont quasi inexistants dans l'habitat standard urbain.
D'un point de vue environnemental, on constate un épuisement des ressources, un accroissement de la pollution et des coûts énergétiques de l'habitat élevé.
Enfin, concernant l'habitat, la qualité du cadre de vie fait défaut. Les logements urbains sont manquants ou trop chers. Le nombre de bidonvilles augmentent.
La question est donc de trouver comment construire des lieux de vie répondant aux envies et aux besoins de tous, tout en préservant la planète et ses ressources, alors que la population urbaine ne cesse de s'accroître ? Comment trouver un équilibre entre les terres agricoles et les espaces urbains ?
Ces questions ne sont pas récentes et des propositions émergent pour essayer à de petites échelles de trouver de nouveaux modes d'habiter. Les écoquartiers, en autre, de plus en plus nombreux en Europe, en sont l'exemple. D'autres essais concrets sont réalisés notamment sur le développement de nouveaux matériaux de construction, de lieux de vie avec des espaces de vie collectif. Il y a également des espaces de réflexion qui sont en mis en place dont les workshops font parties.

Les workshops 2013 : un laboratoire expérimental ouvert aux citoyens grâce à la médiation humaine

En partenariat avec l'Atelier international expérimental pour la Cité Bio-numérique, de nombreuses activités ont été proposées dont l'événement majeur a été un projet « Ville numérique et agriculture urbaine bio-high-tech ». Cette manifestation internationale était composée de cinq workshops en continuité entre le 13 juillet et le 31 août 2013. De 10 jours chacun, ils avaient comme thème : le paramétrique, l'automatique, la mutualisation, l'hybridation et l'autoorganisation. Ces workshops furent encadrés par les architectes, urbanistes, paysagistes Mme Claire Bailly et Monsieur Jean Magerand.
Plus de 99 jeunes professionnels et étudiants (coréens, brésiliens, belges, italiens et français...) ont travaillé sur une problématique prospective : développer l'urbain sans réduire les zones agricoles.
L'objectif pour ces participants était de concevoir une ville à haute densité, à haute adaptabilité et à haute durabilité tout en intégrant une agriculture productive, biologique et high tech dans l'espace urbain. Le tout en utilisant ou imaginant des techniques de pointe.
La spécificité de ce workshop était d'être au cœur de la Cité des sciences avec une volonté forte d'ouverture au public. Pour la médiation, les objectifs étaient multiples : mettre le public au cœur du travail de prospective, permettre de débattre de la ville du futur, du développement et de l'aménagement urbain durable, montrer un travail de recherche en cours de réalisation, diffuser, décloisonner les savoirs et enfin impliquer des jeunes professionnels et des étudiants.
Pour la mise en œuvre pratique de ces objectifs, quatre lieux furent aménagés :

-Le premier était un espace de mini-conférence où quotidiennement des spécialistes venaient présenter leur spécialité au sein de leur champ professionnel et partager leur expérience afin d'enrichir la réflexion.
-Un espace de travail pour les participants où deux visites étaient organisées quotidiennement par des médiateurs pour permettre la rencontre et la discussion entre eux et le public. Un temps primordial pour créer des temps d'échanges entre les habitants et les « futurs bâtisseurs ».
-Un espace dédié à l'exposition de maquettes réalisées sur place, au Fab Lab, outil pragmatique pour illustrer le discours de médiation
-Un espace numérique proposant des tablettes étaient mis à disposition du public pour imaginer et dessiner l'habitat de demain.
Plus de 22000 visiteurs ont pu, grâce à ces activités, comprendre les enjeux de la ville de demain, découvrir les réflexions menés et s'exprimer en dessinant. Pour les participants des workshops, cela leur a permis d'échanger avec le public afin de mieux percevoir leurs doutes et leurs attentes. L'association du grand public dans ce travail de prospective est donc un bénéfice réciproque et permet de bousculer les préjugés et faire évoluer les représentations de modèles urbains.
En concomitance avec ces workshops, l'équipe de médiation a proposé sur cette même thématique un programme d'activités tels que des ateliers scientifiques sur les écoquartiers, l'énergie grise et sur les facteurs qui façonnent l'habitat.

Les workshop internationaux : un travail dans le temps et l'espace

Le workshop international de Séoul (média + architecture) en août 2014 « Bio digital city tomorow's seoul project » s'inscrit dans une continuité et dans une dynamique commune : un travail de prospective sur la ville. La présence de la médiation scientifique de la Cité des sciences a été possible grâce à la collaboration avec L'institut Français.
Le workshop a eu lieu au Seoul Citizen's Hall au sein de la mairie de Séoul, un espace de passage, de rencontre pour les séoulites, au cœur de la cité. Le choix de ce lieu était symboliquement significatif car la participation du citoyen dans la réflexion de l'habitat de demain est fondamentale. L'encadrement par des professionnels internationaux a permis aux participants de bénéficier du retour d'expériences et de la réflexion de leurs pairs mais également d'apporter une dimension culturelle. Celle-ci est d'autant plus renforcée que les participants étaient également de différentes nationalités.
A proximité des espaces de travail du workshop, des œuvres d'art étaient exposées. Ce dispositif permettait d'attirer plus de visiteurs tout en leur donnant un autre regard.
Au fur et à mesure du temps, le dispositif des workshops devient un concept robuste qui s'enrichit au fur et à mesure permettant une progression idéologique et pratique.
Fort de l'expérience de ces workshops, des projets sont en cours, lors de l'année de la Corée en France à la Cité des sciences en 2015 puis à Mons (Belgique), capitale européenne de la culture en février 2015 et enfin à l'occasion de l'année France en Corée en 2016.

by universcience Paris

Universcience is the first French public establishment for the dissemination of scientific and technical culture born out of the merger in 2010 between the Cité des sciences et de l'industrie and the Palais de la découverte. Its 3 million visitors per year have made Universcience the 4th most visited cultural site in France and the leading science centre in Europe. Its interactive and fun approach is aimed at tackling science and techniques through the discovery of the major challenges facing our world. Its offers range from temporary or permanent exhibitions, spaces dedicated to children, resource facilities, experimentation with innovative digital tools.

Universcience shares its know-how in France and internationally with a view to spreading scientific and technical culture.

• touring exhibitions, from 50 sq.m to 700 sq.m, for all audiences and covering a wide range of topics;
• educational spaces dedicated to children, "Cité des enfants";
• cultural engineering and museum consulting.

Universcience est le premier établissement public français pour la diffusion de la culture scientifique et technique né de la fusion en 2010 entre la Cité des sciences et de l'industrie et le Palais de la découverte.
Universcience est le 4ème site culturel le plus visité de France avec 3 millions de visiteurs par an et le premier centre des sciences en Europe. Sa mission est aussi simple que sa réalisation est ambitieuse : transmettre au plus grand nombre, l'envie de comprendre le monde qui l'entoure, permettre à la fois l'émerveillement devant la science et le questionnement scientifique. Ses offres vont des expositions temporaires ou permanentes, des centres de ressources, d'expérimentation avec des outils numériques innovants.

Universcience fait part son savoir-faire en France et à l'étranger en vue de diffuser la culture scientifique et technique :

• expositions itinérantes, de 50 m² à 700 m², pour tous les publics et couvrant un large éventail de sujets
• espaces éducatifs dédiés aux enfants, "Cité des enfants"
• expertise en ingénierie culturelle

BIO-DIGITAL CITY WORKSHOP
11 PROJECTS

EDIT
BAEK Seungman, Jean MAGERAND, Claire BAILLY

DESIGN
SHIN Hanna, NAM Youngju

TRANSLATION
PLC works

SUPERVISION
Stephane LANTEIGNE, Emilien GOHAUD, BAEK Seungman

PRODUCTION
Daewondang design & printing

First Published in November, 2015
By Yeungnam University press

ISBN 978-89-7581-515-7
20,000 won

Printed in South Korea